Management for Professionals

More information about this series at http://www.springer.com/series/10101

Theo Janssen

Enterprise Engineering

Sustained Improvement of Organizations

Springer

Theo Janssen
THR Janssen Beheer BV
Wijchen, The Netherlands

ISSN 2192-8096 ISSN 2192-810X (electronic)
Management for Professionals
ISBN 978-3-319-24170-8 ISBN 978-3-319-24172-2 (eBook)
DOI 10.1007/978-3-319-24172-2

Library of Congress Control Number: 2015954972

Springer Cham Heidelberg New York Dordrecht London

Printed on acid-free paper

Springer International Publishing AG Switzerland is part of Springer Science+Business Media (www.springer.com)

Introduction

It must have happened some 20 years ago, but I remember it well. I started my working life working for a temporary employment agency and I was stationed to work in an ICT center of a large government institution.

I was fascinated by the fact that employees could be found discussing "the organization" at different times of the day. The organization was often referred to in the third person, and mostly negatively—as if it were a sort of monster that wandered through the building and could attack them at any moment. Everyone was bothered by this "monster," but found it difficult to defend themselves against it; it was just there, and there was always such a resigned atmosphere when employees were discussing it. This monster made its presence known in different ways. For example, people would say: "I don't know who is responsible for that." Or it was established that "those others from department X want to do this work, but it is already our job." And it went on and on: "I don't know who manages that," "we have made agreements with that other department, but they are just doing it their own way," "That is just how the organization works," and "That is not our responsibility." These conversations about problems concerning the cooperation and responsibilities/authority could drag on and on, could occur at any moment of the day, and negatively affected productivity.

When I would ask what caused these types of problems, the answer would always be "the organization." When I would ask what they meant by that exactly, I was shown an organizational chart. Apparently the monster looked like a rake, which attacked employees with its sharp points the moment they were just about to become productive. It was a strange sensation. Still, I did not dwell too long on that phenomenon at that time; after all, I had just started working, and I put it down to my lack of experience. After having worked at the ICT center I was positioned in many other large to medium-sized companies and was surprised to find that the same problems arose everywhere. That meant that what I had experienced in my first job was everything but unique. On the contrary, it was a common occurrence. Much time was wasted on discussions about who should do what and who was responsible and there were "fights" between departments about the tasks to be performed.

What an enormous waste. After all, cooperation can take one to great heights. It is the foundation on which civilizations were built and that enables societies to continue functioning, just look at the famous examples in the history books.

For example Caesar's victory on the Gaul at Alesia (which is assumed to be the modern-day Alise-Sainte-Reine, northwest of Dijon). The Romans were by far in the minority, but succeeded in subduing the Gaul by simply working so well together when performing their combat techniques and their disciplined interaction. There are also some good nonmilitary examples of what a good cooperation can achieve. If people don't work together, we achieve little. It just goes to show how important cooperation is within an organization, as it allows it to run smoothly.

Later, when I had set up my own ICT consultancy firm, which also focused on the performance of change programs, I was faced with the organization monster from a different perspective. The key question in change management is how to involve people in the change, and how to overcome the resistance. The libraries are filled with books on this subject. My employees and I are well versed on this subject and had designed an approach that would tackle all these problems. We called this the "middle-out" approach, i.e., the management dictates the boundaries in which change can take place, after which the employees initiate the change. We initiated and guided this process. Our clients were enthusiastic about the visible effects of this approach and we thought we had tamed the organization monster once and for all.

However, we returned to the client a couple of years later, only to discover that little to nothing was left of the change initiated at the time. We were surprised to find that the client was still very satisfied, despite the lack of result, welcomed us as their great savior, and asked us to set up a new program. We asked them why they were still so happy about us, and they responded by saying that our approach back then was excellent.

After all, they considered us not to blame for not having sustained the results of the change we introduced a couple of years before. They confirmed that a new change program had been initiated after we had left or sometimes even a complete reorganization, which we could do nothing about.

Call me an idealist, but I was not happy to hear this. We came there with the intention of implementing permanent change, and we had not succeeded in doing so. We had only tamed the organization monster for a limited period, and as soon as we had left, it reappeared and destroyed all the results. I couldn't accept that. How was that possible? We thought we had executed the change measures correctly. We paid attention to the human aspect and we had introduced real changes in behavior. How was it possible that hardly a trace of that change was to be found afterwards, not in the workplace, nor in the management?

I decided to study organizational behavior and corporate science even closer. I wanted to find out how it was possible that organizations always resort back to the performance level as it was before the change. According to the doctrine of organizational behavior and corporate science, you should, or so I read, be able to manage an organization as a black box: we don't know exactly what happens in that organization (the "black box"), but you can still change the organization just by pushing a number of "buttons" and then measuring the effect this has had. The idea of seeing an organization as a black box was explained by the organization being a complex system: too complex to visualize its inner workings. The various elements contained in that system respond to the adjustments made here and there, eventually leading to a certain result. But is that also the desired result? In light of reorganizations, mergers, and failing change programs, I felt the essential question was: can an organization be managed as a black box, just like for example a car, which most people seem to be able to drive without having even the slightest bit of knowledge of car engineering? However, unlike cars (which are extensively tested before they can be sold), organizations are never tested on their internal operations before they are set free into the world. It is therefore not so strange that organizations are so difficult to manage and change.

The reason why changes are often not successfully sustained in practice is because we often don't focus on the way we work together, i.e., the specific activities that make up a cooperation pattern. For example something as simple as the weekly collection of garbage. You can consider that a service provided by the municipality in exchange for the taxes you pay as a resident. But, of course, things are not that simple in practice, because as a resident, you are required to deposit your garbage in a plastic bag or roll container and to ensure that this bag or container is located by the side of the road (and for example not *earlier* than a certain time the evening before). The activities <resident puts garbage in bag>, <resident places the bag by the road on day x before u hour>, and <municipality service (or service provider assigned by the municipality) collects the bag on day x between u and y hour> are all elements of the cooperation between the resident and the garbage collection service of the municipality.

Everyone surely knows from their own experience that things can go wrong with the collection of garbage. You forget to put your bag or container on the street, you have developed a habit to do it in the course of the morning on day x, as the garbage truck won't come by until noon that day. You are annoyed if you are unlucky enough to see him driving through your street at 9 a.m. In that case both the resident and the garbage collection service have filled in their mutual cooperation differently, without first discussing it with their cooperation partner. As a result the cooperation fails: The garbage truck cannot collect the garbage, and the resident is left with his garbage.

This is just a simple scenario of the cooperation between only two parties (*one* family in relation to *one* municipal service which is provided *once* a week). Just imagine *the* cooperation in *the* organization: this involves an endless amount of relations and interactions in which anything could go wrong. If we look at it this way, it is a miracle that so much still goes right in society and in organizations.

This, in a nutshell, is what Enterprise Engineering is about: mapping the crucial agreements and activities that make up a cooperation pattern, and research its possible flaws. In 2009, I was introduced to this area of expertise, which is completely new and that, contrary to many other methods, is based on thorough scientific research by, for example, the Technical University of Delft. And what did this research claim? That it is possible to study, describe, and change organizations as a "white box," i.e., a fully transparent organization, just as this is possible for a physical construction of, for example, a car. If that is true, I thought, then that is the method I have been looking for—something that would take care of that organization monster once and for all. I have studied this area of expertise and have spent 2 years translating the theory into practical applications. I managed to complete my work and the results are extraordinary. The experiences with Enterprise Engineering in practice, many examples of which are included in this book, show that if we adopt this principle and look at the cooperation and therefore organizations in this way, changes actually take root and improve organizations on a structural basis.

It is not particularly easy to familiarize oneself in the field of Enterprise Engineering. It is a very complex study, which explains why this method hasn't yet been fully accepted by companies and society. However, the value of the method is so great that I felt that there had to be a book that could make this unique field of expertise accessible to a wider audience. That is the book you are currently holding.

We will see that an organization consists of elements other than departments, functions, and processes. We can clearly dissect the cooperation which is the very heart of the process. Only when that has fully been uncovered, do we see where "the organization" is at fault. We need that insight in order to know which changes are required in which area. As a result work can be performed much more efficiently in an organization. This not only benefits the cooperation, but can also result in substantial cost savings. In addition, it makes the organization easier to manage and allows it to be changed more easily, for example, if a new strategy would require so.

It is not my aim to provide a scientific justification for Enterprise Engineering. If you are interested in learning more about the academic background and extensive methodical principles, I gladly refer you to the excellent book *Enterprise Ontology* by Jan Dietz [1].

This book allows you to look at organizations in a whole new way. It also provides ways of using that method. The cooperation in the complex system of an organization will no longer be a mystery, but a clear and usable phenomenon that can be adjusted to meet the requirements of the board. The organization is no longer an untamable monster: a true change of paradigm.

Wijchen
The Netherlands, July 2015

Theo Janssen

Reference

1. Dietz, J. L. G. (2006). *Enterprise ontology. Theory and methodology*. Berlin: Springer.

Word of Thanks

My whole life I tend to deviate from the conventional path and pave new roads for others for the purpose of making the world just that little bit better. My focus therefore always lies on innovation. "Good" is not good enough and even the enemy of "better." It does not make my life any easier. Ten years ago, I knew everything about "lean" and "learning organizations," something everyone is following at the moment. Now I have realized, in practice, that the return of these methods is too low, certainly in the long term. And I know exactly why it goes wrong. But it is not done to say something negative about these subjects, as the hype is unstoppable, at least for the moment.

That is why I am thrilled that visionaries at companies have given me the chance to show that Enterprise Engineering is able to bring about real and sustainable changes in organizations. I am particularly grateful to Rob Bakker, Erik Passchier, Adri Risseeuw, Wim Uittenhout, John Heideman, Eise Douma, and Steven Wauters for having faith in me and in Enterprise Engineering.

I would like to thank the dedicated employees of Delta Change Consultants for the discussions and establishing the list of Elementary Construction Flaws and the clear models for the cases.

Joep Smaling provided an important editorial contribution to this book. Without him, I wouldn't have succeeded in shedding the scientific terminology of this subject and express the message in clear language. Peter Brouwer helped me substantially in reviewing both the Dutch and the English texts. It's amazing to see how many errors translators make which Peter can detect and correct.

And finally, I would like to express my thanks to my dear wife Wendy and our beautiful daughter Phoebe. In addition to my busy work schedule, I decided to write a book. They have supported me during this process, even if it meant they saw less of me.

Contents

List of Figures

Is There Any Use in Changing? 1

There are quite a few books for sale on organizational behavior in general, and management of change in particular. One would expect that those trees were cut for a good reason—in short, that we would know by now what the best change program and what its best implementation strategy is. But sadly more than 70 % of the change programs worldwide do not achieve the desired return [1–3]. This percentage is even higher according to some researchers, and there are even researchers who claim that not a single change program has had the desired effect [4]. Also 96 % of all the mergers and takeovers are very problematic (research Hewitt and Associates, 2009) apparently as a result of 'cultural problems'. These are rather shocking findings, and one thing is for sure: there is something wrong despite the willingness of all the parties involved.

1.1 Six of One Half a Dozen of the Other

The progress of a change program is predictable: the organization often benefits right after the changes have been implemented. But eventually this wears off until the changes and the goals made in this respect have all disappeared. The top then usually decides to implement a new change program that will then disappear in the same pattern. That is not only a pity because changes can bring a lot of good, but also painful in view of the time, money and energy invested for nothing.

A second aspect of the problem with change programs is that the organization cannot really be changed to meet the wishes of, for example, the board. By 'really' I mean that the organization would exactly undergo the changes as desired. After completing the program, it may become clear that the changes made in the organization bring with them some unpredicted and undesirable effects.

And to make matters even more complex, organizations tend to automate many of their processes. All the effort required to achieve optimum alignment of business operations and ICT is referred to as 'business-ICT alignment'. It reveals that ICT systems do not or rarely operate as expected. This not only dissatisfies the users of

T. Janssen, *Enterprise Engineering*, Management for Professionals,
DOI 10.1007/978-3-319-24172-2_1

the ICT systems, but also those higher up in the organization will raise the question as to whether ICT services actually support the company's strategy. The return on investment promised by ICT is often not achieved.

1.2 What Is Real Change?

The problem is not that changes are by definition bad. The problem is that one applies changes without first studying the *heart* of the organization.

By this I mean that neither managers nor change experts see what is 'wrong' with the organization exactly. They do have an idea, because they wish to change 'something'. And they have (probably) also agreed on that. But this does not mean that their conclusions on the true nature of what is 'wrong' are also correct. Enterprise Engineering, which we will see later in this book, helps track the real problems, as opposed to the problems that are assumed to exist.

The issue of this problem of failed change programs and reorganizations has been tackled by ICT scientists—exactly because they were often faced with issues aligning ICT to the business operations. Change experts and organizational experts continued to fail; they failed to take the initiative to thoroughly investigate these issues from a scientific perspective. They did suggest so-called 'solutions', but did not determine the actual causes. The founders of Enterprise Engineering discovered that not only was there something wrong with the communication with and within ICT systems, but also with and within organizations being overall systems.

Organizations have been regarded as systems for decades for all the good reasons. They are social systems that only exist and function because they are made up by people and the way people work together. Without that cooperation nothing will come about. If we wish to research organizations thoroughly we will therefore need to take a closer look at the cooperation. But what does cooperation look like exactly? It appears that the cooperation can be researched and described in detail, much more detailed than, for example, a description in terms of processes.

The cooperation takes shape via a universal pattern which is referred to as a 'transaction' in Enterprise Engineering. It is the way in which two people reach an agreement, achieving a certain result. That result does not directly constitute the final product provided by an organization, for example a mortgage for a bank, or a bread for a bakery. Before the final result is achieved many more agreements need to be made, each having a partial result. A bakery must, for example, first order the ingredients. This requires him to make arrangements with someone else, the result of which are the delivered ingredients.

All those agreements, results and the way in which agreements and results are linked form the *construction* of an organization. The construction of an organization can be revealed, and made as clear as, for example, the construction of a car or a house. And as we will see, this construction is also the heart of an organization.

But what use is it? And does this guarantee that change programs will no longer fail? There is never a guarantee of success, but one thing is for sure: a bad solution results in failure, guaranteed.

That's why we should choose a good solution. Once the construction of an organization has been revealed, we have a detailed overview of the cooperation and therefore also of what goes wrong. Without this analysis, the organization fails to utilize many possibilities for improvement. And what's even more valuable: the construction can now be changed to suit the requirements of the user. Any changes made will be implemented structurally in the core of the organization; they are no longer 'covering' it, like a blanket which you can easily throw off.

However, Enterprise Engineering is not a method for determining the best company strategy. That is a totally different ball game and at least as important as a good cooperation. An Enterprise Engineer works on the premise that a strategy has already been adopted. Still an organization that wishes to introduce a specific strategy can't really do so without Enterprise Engineering, as this is the ideal method of finding out whether the organization can handle the strategy in the first place. An organization can avoid a lengthy and costly process of trial and error if they would first examine its inner workings before implementing a strategy. But I will discuss that later in this book.

So yes, changing does have its uses, but not in the old way, not on the basis of untested assumptions and vague solutions, but rather on the objectively determined construction of an organization.

References

1. Bower, J. L. (2000). The purpose of change: A commentary on Jensen and Senge. In M. Beer & N. Nohria (Eds.), *Breaking the code of change*. Boston: Harvard Business School Press.
2. Boonstra, J. J. (2000). *Lopen over water: over dynamiek van organiseren, vernieuwen en leren*. Amsterdam: Vossiuspers AUP.
3. Balogun, J., & Johnson, G. (2005). From intended strategies to unintended outcomes: The impact of change recipient sensemaking. *Organization Studies, 26*, 1573–1601.
4. Homan, T. H. (2013). *Het et-cetera-principe – Een nieuw perspectief op organisatieontwikkeling*. Den Haag: Academic Service.

The Traditional View on Organizations 2

In the introduction I referred to Enterprise Engineering as a paradigm change. A paradigm is the reference framework from which we interpret reality. Enterprise Engineering allows us to change the reference framework and look at organizations in a whole different way. But before we look further into the new paradigm, we first need to establish what our old reference framework was. We will therefore discuss the current approach in this chapter, whereby the organization is seen as a black box. Many managers, experts on change and others involved in the functioning of organizations often don't realize they look at organizations in this way. However it is clear from their actions that they do.

2.1 The Organization as Black Box

The metaphor of the black box relates to systems. The qualification 'black box' is used to typify a system that is not known: the exterior is visible, but not a lot more. Something goes in and something comes out, but what really goes on in that system is unknown, as it is invisible. This is probably the main reason why there are so many different views on organizations. The history of organization science has seen the popularity of one vision, followed by another, but never did they establish an 'ultimate' version. The different views on organizations can be regarded as push buttons on the black box. Each organizational expert has his own ideas of the right setting. He will for example—and I am simplifying matters—push a certain button fully or will try and use a combination of various buttons in order to achieve the effect he is seeking. A bit of this and a bit of that: one expert might suggest that his vision will allow the organization to be more cost efficient by applying a certain method to change the processes. The other might recommend, depending on the final goal of the change, to do something with human resources and to implement self-managing teams. The expectation is then that such interventions should have effect.

T. Janssen, *Enterprise Engineering*, Management for Professionals,
DOI 10.1007/978-3-319-24172-2_2

Initially there is indeed some effect. If, for example, staff is dismissed, then this reduces costs. But note: this is a black box method, based on expectation and hypothesis. How can management be *certain* that a specific employee in a specific position is important for the organization or not? The answer is simple: they cannot. They think they know, they have a feeling and arguments will be made of the supposed 'redundancy' or 'dispensability' of the employee in question. Maybe it will eventually become clear that the right choice has been made, and the employee reduction will not have a negative long-term effect. But as the black box does not allow for any insights, it could just as easily become clear later that a very important position has been given up. A certain type of work that is of great importance to the department goals, and therefore the organization goals, is no longer being performed. Dismissing employees is of course just one of the examples of a measure that is performed 'in the dark' as it were. In fact the management is steering 'blindly' in respect of each measure and decisions are not based on a thorough and objective knowledge of the organization.

There is some pushing and shoving, there is some effect, but the effect turns out to be unwanted and/or seems to generate a lot of new problems.

This old paradigm is still dominant. Even though many organizational experts think they can look under the 'hood' of the organization, for example by using a root-cause analysis, they can't. They lack the right tools: a method which will show the construction of the organization, i.e. the inside of the black box.

Without the knowledge of the internal operations changes are pushed forward and forced upon the organization.

Everyone understands that this does not work on physical constructions such as cars or houses. If a BMW is broken, the mechanic will check under the hood to establish a diagnosis. And if for example the fuel injection has broken, he will determine whether the broken part can be repaired or needs replacing. Naturally he will not touch the parts that did not cause the problem. He will not pull on the window wipers or install a new exhaust. And he will find a suitable replacement part, a part that fits the construction of the specific BMW.

He will not try and fit Toyota parts into a BMW. The construction of the fuel injection of a Toyota does not fit into the construction of a BMW. Everybody understands that.

But in organizations, people still tend to operate 'blindly'. One does not know the construction and can therefore not determine an objective diagnosis. Are we actually dealing with the part that really needs replacing? Your guess is as good as mine. Are we going to fully replace the part or just amend it? And why replace it all if that is the decision? Does the new part fit in the first place? Well, who's to say? We will probably find out in 6 months.

And then it becomes clear that the outcome is undesired.

Change program follows change program, adjustment follows adjustment without achieving the desired sustainable changes. The only thing that changes 'sustainably' is that the budget is exceeded time and again.

2.2 Big Organizations, Big Problems

Those who think that these problems only happen in small organizations are mistaken. It happens everywhere. These are just some examples.

2.2.1 ABN AMRO: The Efficiency Ratio

In his bestseller *De Prooi* (The Prey), author Jeroen Smit describes why ABN AMRO (a large bank operating worldwide and established in the Netherlands) went under in its original form as hunter and eventually was cut up as prey and sold to three foreign banks [1]. The book deals with different subjects. One of these subjects is the efficiency ratio. For years the bank tried to improve this ratio representing the costs as percentage of the turnover: the lower this ratio, the more efficiently the bank was able to operate. The efficiency ratio of ABN AMRO was always much higher than desired, it fluctuated around 70 %. Naturally the shareholders were not happy about this, because a high efficiency ratio means that a lot of profit evaporates, which for them meant: less profit per share. The shareholders put pressure on the Board of Directors to find a solution for the high efficiency ratio. *De Prooi* covers a period of about 25 years, during which much was tried to make the organization run more efficiently. A lot of investments and efforts went into the preparation of a merger. Then there were cuts in spending, people were laid off and exorbitant bonuses were dealt with. And after a while attempts were made to change the corporate culture. They experimented with personal coaching, group coaching and company trips with fun activities to bring people closer together. Nothing helped. The efficiency ratio remained too high, seemed to fluctuate at random, and sometimes went down by a couple of surprising percentile points, followed by an increase. Millions went into fees for consultancy firms, one even more expensive than the other, to introduce changes that eventually resulted in nothing. Nothing? Actually, there was a result: the bank was taken over and the rest is history.

2.2.2 Integral Care

In the care sector, there is more and more cooperation. Various care providers want to fully cooperate in order to improve the quality of care by good coordination. Naturally economies of scale play a role as well. Mergers and other types of collaboration are practically an everyday occurrence.

Sadly the cooperation between some institutions leaves a lot to be desired. A good example is what happened with IJsselmeer hospitals. The two hospital locations in the Netherlands, Emmeloord and Lelystad, merged in 1990 and it immediately became clear there were all sorts of problems that continued for years. The organizational chaos resulted in a quality problem and the Health Inspection had to step in. Eventually the Inspection even went as far as imposing a ban on performing surgery [2]. In 2009 IJsselmeer hospitals were on the brink of bankruptcy, but eventually the institution was saved by the government and other parties. The main cause of the problems was said to be the bad cooperation between the board of directors and the supervisory board. The board of directors, in particular, were said to have no idea what exactly was going on in the organization. An organization is a cooperation network, so naturally the problem must concern the cooperation. But who's to say what the real cause of the cooperation issues at IJsselmeer hospitals was?

As long as the problem is not studied by using a proven objective method, all the patterns of cooperation, including what goes wrong, remain unclear. The affairs surrounding the Academic Hospital of the VU University in Amsterdam, VUmc, are also poignant examples of cooperation problems. In 2012, the whole hospital was placed under increased surveillance by the IGZ (inspection authority for healthcare) and the lung unit was closed after evidence of serious misconduct had been established at the intensive care unit a year before [3]. An official statement was made that there were 'social problems' between various specialists.

Cooperating appears to be not that easy. When it involves relatively simple tasks in small settings, it is not so difficult. But the cooperation within and between complex organizations is often below the desired level.

2.2.3 Defense and ERP

Enterprise Resource Planning (ERP) software programs are used to support all sorts of business processes. Various separate applications, for example for the management of client data, financial administration and stock level data, have been replaced by ERP packages. These ERP packages for example use one single database, which means data needs to be entered only once in one overall system that links the data. The people using the system only want one thing: that the system does what it is supposed to. But that's exactly where it almost always goes wrong.

In 2005, the Ministry of Defense of the Netherlands decided to purchase an enormous ERP system from the company SAP and to implement it throughout the organization. This involved a total restructuring of the ICT landscape. Things went rapidly downhill from there: there were 'start-up issues', difficulties with the planning, goals were not met, and the budget was exceeded by over 80 million euros. Of course there was lots of speculation as to the causes: cultural differences between the divisions of the Defense organization, the budget was too tight, unclear additional terms and conditions et cetera. Is something wrong with ERP as a system? No, a total package such as ERP has many advantages and the people who made it are usually experts. But there is a knowledge gap, the people making the package are not aware that an ERP package also houses a construction and that the organization where the package will be used, also has a construction. The chances of these constructions being compatible are negligible. If one had taken the time to analyze the cooperation within the organization and had made sure that it seamlessly suited the cooperation prescribed by an ERP package, then one would have had a chance that the package would actually deliver the promise. However, the 'plug' of the package did not fit the 'socket' of Defense. This meant that the implementation at the Defense department eventually had to be aborted, resulting in a disinvestment of a billion euro [4].

Whether it is about changes in organizations, such as ABN AMRO, or about the cooperation between organizations in the healthcare sector or ICT system related changes, problems related to change are widespread. However large, rich, famous or professional the organization may be.

The experts setting to work on the change—often reputable agencies—are knowledgeable, but unfortunately lack the right tools to discover what's in the black box.

2.3 A Brief Outline of the Old Paradigm

It is not surprising that we still try to fix organizations by trial and error as there so many different views and subsequent methods of change can be found in the history of organization science. The theorists did aid the advance of organization science with these views. Much thought has gone into the various aspects of an organization, the people, the processes, management and change. But none of these views has provided an insight in the black box. Most universities and colleges are still not aware that Enterprise Engineering has brought about a change. This paragraph provides a short outline of the main movements to illustrate that, however much they differ, they fail to include the construction of an overall organization.

Organization science, also referred to as organization theory, has become a formal science after the Industrial Revolution took off. The early years of organization science—historians have established that the first period is from the end of the nineteenth century until just before the Second World War—were largely dominated by the ideas of Frederick Taylor (1856–1915). This period is also referred to as the period of *scientific management*. Taylor mainly drew up scientific observations of manufacturing departments, for the purposes of making analyses that would increase the efficiency, in which he succeeded. This resulted in stricter division of work: each employee was given his or her own task. This meant the end of the traditional way of thinking and acting based on craftsmanship. The rationalization of the work method was to take precedence.

The mining engineer Henri Fayol (1841–1925) was mostly involved in the skills required for the management of an organization. He suggested a strongly normative general management theory, usually referred to as the *general management theory*. He distinguished various organization functions and determined what he considered to be the five most important management tasks, his five pillars of modern management: planning, organizing, commanding, coordinating and controlling. He also established management principles, such as the 'unit of command'-idea. This meant that each employee only had one manager. This resulted in a chain from the top to the work floor, the structure that indicates the direction of the commands. Also the functional set-up of departments, teams et cetera was strongly emphasized. Fayol,

Max Weber and Taylor were the founders of what is now referred to as the *classic management theory*. Weber, a pioneer in sociology, paid a lot of attention to the acting based on understanding its purpose and the ideal shape of bureaucracy as a straight, honest and effective management method.

Organizations set up according to these classic management principles were very hierarchical, with strict systems and procedures based on military principles. Salary was seen as the only way of maximizing productivity.

More recent experiments, such as the famous Hawthorne experiments in the 1930s lead to new insights [5]. They made it clear that labor productivity increased if more attention was given and interest was paid to the employee as a person. Mutual solidarity also appeared to have a positive influence. Management thinking paid attention to what was described as human relations: the relations between people within organizations. The classic management theory disappeared into the background. Just about each decade saw the introduction, merger or total dismissal of theories.

More emphasis has been placed on organizational processes in recent years. A process is defined as a chain of goal-oriented activities with a start and finish, which follow a logical sequence and (should) add value. It was discovered that organizational processes are not affected by the hierarchy within an organization, as they cross department boundaries. Total Quality Management, Lean Manufacturing, Business Process Management are all movements that each aim to manage the complex forces in an organization.

In the view adopted in Taylor's scientific management, the organization was rather simplistically considered a machine. Naturally this was a misconception. As we were to discover later, organizations are complex to an extent that all the nuances would be lost if such a dogmatic approach were to be applied. The movements that followed scientific management therefore took into account that the organization is so complex that it would be best to consider it a black box. This also turned out to be incorrect, as it has now become clear that this complexity can be fully revealed.

But let's not get ahead of ourselves. There is also a group who considers that the black box can be changed into a white box, i.e. a fully transparent organization, by applying the process-oriented approach. But they too are wrong, as they fail to understand the organization at a deeper level.

It is a common misconception that Enterprise Engineering is just a new way of describing processes. It is a way of revealing the deepest layer of the processes and therefore the deepest layer of the organization.

And finally a remark on the more modern psychosocial approach. The behavior of managers and other employees is often based on expectations, assumptions, values, thoughts and feelings. They are people of flesh and blood, not robots who act purely rational. That is why organization science also pays attention to how employees and managers are optimally valued and motivated. Empowerment has become a well-known phenomenon: you should empower employees as much and as smart as possible to get the best results. Provide them with flexible work areas, a company bike and a first class lunch. It is as if organizations are looking for ways to comfort their employees because they are so frustrated with the daily course of events. This is also a completely subjective approach. Who can objectively determine that employees become intrinsically more motivated or loyal as a result of this approach, also in the long term? And that this always leads to a better organization?

If we really wish to change an organization for the better, we need to look beyond the exterior if we are to move forward. We can distinguish numerous aspects and make useful comments about them, but real progress is only made if it irrefutably determined what actually takes place *within* the system. We must therefore visualize the system, or lift the 'hood'. Then we don't have to fumble in the dark anymore and we will be able to see what needs adjusting and the consequences this will have.

References

1. Smit, J. (2008). *De Prooi*. Amsterdam: Prometheus.
2. Van Kemenade, L. (2008). *IJsselmeerziekenhuizen krijgen operatieverbod*. Elsevier.nl. http://www.elsevier.nl/Stijl/nieuws/2008/9/IJsselmeerziekenhuizen-krijgen-operatieverbod-ELSEVIER203816W
3. ANP. (2012). *VUmc onder verscherpt toezicht IGZ, longchirurgie tijdelijk gestaakt*. Trouw.nl. http://www.trouw.nl/tr/nl/4516/Gezondheid/article/detail/3304277/2012/08/21/VUmc-onder-verscherpt-toezicht-IGZ-longchirurgie-tijdelijk-gestaakt.dhtml
4. Dijkhuizen, B. (2014). *ERP-project Speer kost Defensie 900 miljoen euro*. Logistiek.nl. http://www.logistiek.nl/Supply-Chain/erp/2014/1/ERP-project-Speer-kost-Defensie-900-miljoen-euro-1447420W/
5. Franke, R. H., & Kaul, J. D. (1978). The Hawthorne experiments: First statistical interpretation. *American Sociological Review, 43*, 623–643.

3 A New Paradigm

New paradigms don't just fall from the sky. They are the result of different development in various areas, science being the most important one. The most famous example is that of Galileo; Galileo defended Copernicus' theory that the Earth revolves around the Sun and is not the center of the universe. The new paradigm we will discuss in this chapter is maybe not as great, but does have substantial consequences for organizational science. But before discussing this new paradigm, we will first take a look at the three important scientific theories on which Enterprise Engineering is based.

3.1 System Thinking: Hidden Connections Uncovered

Hundreds of books have been written about systems, and there are hundreds of definitions of what a system exactly is. Consensus has been reached on the four main characteristics of a system. The system has:

Composition: the system consists of various elements
Structure: the elements in the system affect each other
Production: the system generates goods or services
Environment: there is a clear boundary between the system and the environment.

Figure 3.1 shows how elements in a system influence each other and how these elements can be connected to the environment.

The concept 'system' and the 'systemic' way of thinking, also referred to as 'system thinking', are important. It strongly emphasizes the relations between the elements. A system is a system because it is formed by elements that are *somehow* related. A human organism is a system that consists of subsystems, such as a subsystem for burning fat and a subsystem for digestion. The subsystems of a car are the driving mechanism and the brakes. An organization is a system too and consists of sub-systems.

T. Janssen, *Enterprise Engineering*, Management for Professionals,
DOI 10.1007/978-3-319-24172-2_3

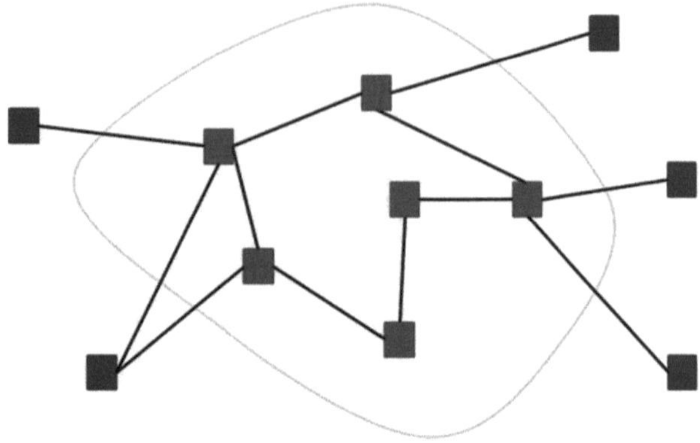

Fig. 3.1 All elements of a system

System thinking enables us to think about elements in a structured way, and about where and how they are related; for example in the form of mutual dependence and interference.

Peter Senge placed system thinking on the map with his popular book *The Fifth Discipline—The Art and Practice of the Learning Organization* [1]. Senge is a great believer of the learning organization, but he claims that the reason that organizations don't learn is because they have 'learning disabilities'. In other words: they keep on making the same mistakes. Not because the people working in those organizations are incompetent, but because they are placed in structures where the same mistakes are repeated over and over. According to Senge, the problems experienced today are caused by 'solutions' of the past. This has to do with the dominant and incorrect idea that there is a linear—direct—relation between cause and effect that are in close proximity of each other, in terms of time and space.

For example, if there is a problem with customer satisfaction, all the employees can be sent to attend a course to provide a better service to customers. The assumption seems to be that the problem is caused by the attitude of the employees. This is not system thinking, but thinking in terms of cause and effect in a linear sense: the customer is dissatisfied—the cause must be that the customer is treated badly—who treats the customer?—the employee—therefore: the employees must attend a course to learn how to provide a better service to the customer.

The assumption is that cause and effect are in close proximity of each other in terms of time, but nobody knows for sure. However, it is equally possible that activities are planned in such a way that employees are given too little time to resolve the problems of the customer. This would mean that the problem has a completely different cause than the employee's attitude towards the customer, but nobody sees this.

Another example that provides a clear illustration of cause and effect not always being in close proximity of each other in terms of time, can be found in the example of ABN AMRO we discussed earlier. There we saw that the shareholders considered the efficiency ratio too high. All sorts of remedies were tried in order to reduce that ratio. One year they succeeded: the ratio had fallen by two percentile points.

Did this mean that the system could be influenced externally after all? No, sadly not. The year after the ratio had again risen by four percentile points. It turned out that the success achieved was just temporary. As a result of the implemented measures the employees who were responsible for selling the bank products were feeling less comfortable and became less and less motivated to do a good job. This caused the turnover to drop and the ratio to rise again. These additional effects of changes need time to develop. In his book Senge refers to this as 'delays'. This linear way of thinking doesn't do justice to the complexity of systems. In order to find fundamental solutions for problems, one needs to have a thorough understanding of the system and change it where necessary. System thinking helps people comprehend and see through the world.

3.2 An Organization Is a Social System

Sociologists, psychologists and many others in the social sciences have also studied systems by viewing organizations as systems. Organizations as *social* systems to be precise. An organization only exists because different people work together in a formalized situation. They also work effectively and make arrangements with each other, there is no organization without these people. Desks, chairs, laptops, paper, organizational charts, process descriptions et cetera. will do nothing on their own. *People* set things in motion, make arrangements to produce something together, and *people* take the decisions.

In a technical system the elements that form a system often have no other choice than to work together. They are in a fixed location in the system and have no choice in the matter—the pedals of a bike drive the gear wheel, which drives the chain et cetera. In a social system the mutual interference is naturally more complex and elements, i.e. people, are able to be more flexible in the performance of their tasks in the system. The cooperation between people is brought about by communication. In order to thoroughly understand a social system, we need to look at the way of communication, and the processes used for this communication.

3.3 Language Is the Thread of the Organization

Linguistic pragmatism is a science that researches how language is *used*. Important (language) philosophers who have been active in this field are for example Wittgenstein, Austin, Searle and Habermas [2]. The science discipline based on the work of these philosophers is currently referred to as the *Language Action Perspective*. One of the most important discoveries of the linguistic pragmatism is that *saying* something also means *doing* something. On the one hand a speaker will simply provide information. For example: 'Would you like to go out for dinner with me?' or:

'The menu of restaurant X is exceptional.' The information provided here is: the speaker wishes to take person A out to dinner, and would prefer doing so in restaurant X. But something is also done: in the first sentence a request is 'delivered', in the second sentence a recommendation is given. This 'requesting' and 'recommending' are also referred to as language actions.

Other language actions are for example: promising, stating, accepting, rejecting, warning et cetera. Hence, in order to systematically study communication, these language actions must be taken into account. By determining the type of language action, the expressed communication can also be systematically categorized. There are after all only a limited number of language actions.

3.4 The Rise of Enterprise Engineering

Enterprise Engineering is the discipline that has embraced *system thinking*, *the social sciences* and the *Language Action Perspective* as forming the basis for research to the internal operations of organizations. Enterprise Engineering was first established in the ICT sector. The biggest problem faced by the ICT world for years now, is that ICT systems fail to align with the needs of their users: people. People are social beings and expect an ICT system to deliver something that is apparently difficult to realize. This issue is referred to as *Business-ICT alignment.*

A good example: a former colleague of mine needed to renew his passport at his municipality. He had made an appointment and confidently set about his business. On his way there he realized that his driver's license also needed renewing. No problem, he thought, I will do that as soon as they have arranged my passport. After the nice lady behind the counter renewed his passport, he passed her his driver's license. 'But Sir, we can't do this now,' she answered. 'The ICT system we work with requires us to first plan an appointment with you. Only then will I be able to renew your driver's license. I simply cannot access the right module in the system right now. Are you available in six weeks?' 'Six weeks? But I am here now,' said my former colleague. But there was no other option: the ICT system was set up in such a way that it resulted in situations sketched in that 'purple crocodile' commercial. The designers of the system had not considered the option that a customer might want to perform two transactions, one directly after the other, and therefore this option was not implemented.

We are probably all familiar with that phenomenon. ICT fails to meet humans' expectations. Let's assume this happened at a bakery. You order bread and then tell the baker that you would like a bun as well. 'Come back in six weeks.' I would never come back—what kind of weird bakery is that?! But because there is ICT involved, we just accept it; as ICT never does exactly what we would like it to do. Exactly this situation kindled the development of Enterprise Engineering.

Although the question was first raised within the ICT sector, Enterprise Engineering is not only relevant for the ICT sector. The research was aimed at unraveling the internal operation of organizations. And now it is applicable to all organizations where people cooperate for the purposes of achieving a common goal. A very broad area of application.

As the impulse originated from the ICT sector, most of the practical examples we will discuss in Chap. 7 are situated in the ICT sector. However, these examples will also make clear that the context in which those stories take place is irrelevant. The heart of the story is always about how people work together in organizations. That cooperation always takes place in the same way, regardless of the sector the organization operates in.

A good example of how Enterprise Engineering works can be found in the construction sector. Often dozens of parties need to work together on one construction assignment; not an easy task.

That is why Enterprise Engineering was used to develop a framework to facilitate that cooperation. This framework is called VISI. VISI has been used in thousands of construction projects in the Netherlands and now forms part of an ISO standard (ISO 29481) [3]. VISI is attracting more and more international attention: this framework is exported all over the world. Not so strange if you consider the results: using VISI is advantageous in terms of the quality, efficiency, running time, costs and process management of projects in the construction sector.

For example, the complexity of the cooperation model was reduced by 95 %, archiving activities can be reduced by up to 80 % and the costs of failure (on average 7 % per project) can be reduced by a quarter. See Sect. 7.4 for more details of this case.

References

1. Senge, P. (2006). *The fifth discipline: The art and practice of the learning organization.* New York: Doubleday. Revised and Updated edition.
2. Braet, A. (Ed.). (2000). *Taalbeheersing als communicatiewetenschap. Een overzicht van theorievorming, onderzoek en toepassingen.* Bussum: Coutinho.
3. Jansen, P. (2013). *VISI: communicatie in de bouw.* Crow.nl. http://www.crow.nl/vakgebieden/contracteren/visi-communicatie-in-de-bouw

Under the Hood of the Organization 4

Based on the findings mentioned in the previous chapter and scientific developments associated with them, we can conclude that there is an agreement on the following. An organization is:

a) a system,
b) a specific type of system, namely a social system,
c) and this system functions thanks to defined rules of communication.

These insights have led to the development of a new scientific discipline which helps us understand organizations: Enterprise Engineering. One of the insights that this discipline has added to organizational theory is:

d) this system has a construction that, contrary to what has always been thought, can be made transparent.

To reveal this construction, the cooperation within the organization has to be analyzed. Only then will we discover, firstly, the individual elements which make up the cooperation, and thus, secondly, where the errors in the construction lie. Of course the cooperation within an organization is not a concrete object, like a car. However, like a car, this cooperation comprises several elements which can be isolated, and which may be erroneous. Errors in the construction of the cooperation are called Elementary Construction Flaws (ECFs). These are discussed in Chap. 5.

In this chapter we look at what exactly a construction is and how we can depict it in a model. We will see that such a model provides much more information about the organization (part) than a description of the organization in terms of *structures* and *departments* where people perform certain *functions* when they cooperate.

The concepts that I will introduce in this chapter are new and, therefore, may be difficult to understand immediately. I will explain these concepts by means of an example, in order to make them more understandable for you as a reader. On the basis of this example we will also see how, within organizations, we are exposed to a surplus of information—information that does not actually add anything. The example will show what little of this information really matters: the *essence* of the

T. Janssen, *Enterprise Engineering*, Management for Professionals,
DOI 10.1007/978-3-319-24172-2_4

information in the organization. And that this essence is easier to oversee and to control. First, however, I will describe the example.

4.1 An Example-Case: Maxico

This example describes the complaint handling process at a transport company called Maxico.

4.1.1 Background: Maxico

Maxico, a large transport company, provides public transport by bus and metro in a specific region in the Netherlands. The company is responsible for the transport of more than 50,000 passenger rides per day.

For the registration and handling of complaints, Maxico makes use of the information system COMPLAINTS. There is a process in place for complaint handling in order to resolve disruptions in service as soon as possible. An explicit objective is that the process must handle complaints within a fixed timescale; the exact timescale depends on the priority given to a complaint upon receipt. The organization can choose from five priorities (see Table).

Priority lead time	
1	2 h
2	12 h
3	1 day
4	2 days
5	More than 5 days

Complaints may relate to the transport itself (such as ‘the bus didn’t come’) or, for example, to the supporting ICT systems (‘I cannot recharge my balance on the transport card via the website’).

4.1.2 Description of the Complaint Handling Process

The complaint handling process, as described below, does not yet provide any information about what is essential and what is not. I will demonstrate how, using this process description together with flowcharts, we can separate the ‘wheat from the chaff’ and firmly focus on the information that is crucial for improving the cooperation within the organization.

Complaints about the transport service are filed by e-mail or phone by a 'complainant'. A complainant can be one of the following parties:

- A customer.
- An internal department of Maxico.
- A third party (such as the National Railway Company).

After receipt, the complaint coordinator makes a first scan to check whether the complaint relates to the services of Maxico. The complaint coordinator is the one that follows a complaint from start to finish and ensures that everyone who plays a role in the complaint handling process completes their responsibilities in a timely manner. Both in case of an internal department and a third party, a so-called service agreement is consulted regarding the applicable complaint handling agreements. These may be, for example, agreements related to the complaint handling lead time.

For each complaint received, the complaint coordinator opens a new complaint in the COMPLAINTS information system, and enters the basic facts of the complaint. If the complaint is not the responsibility of Maxico, this will be fed back to the complainant and the complaint will be forwarded to the appropriate party. That information is also recorded in COMPLAINTS, and the complaint is regarded as closed.

If the complaint needs to be resolved by Maxico, based on the information obtained, the complaint is classified by the Service Management department. The classification includes the complaint type. For example, does the complaint relate to service operation, such as a bus that is late, or to bad/undelivered customer service, such as an incorrect travel advice?

Subsequently, Service Management determines the complaint priority, based on the service agreement, the estimated impact and urgency indicated by the complainant. This is done using a formula provided on a Priority Card.

The complaints solution order is determined based on the priorities. With each new complaint received, it is assessed whether a change in the solution order is needed, and whether ongoing activities need to be temporarily set aside in order to give precedence to solutions with a higher priority.

It is Service Management's responsibility to record the complaint as completely as possible in COMPLAINTS, at the same time noting to which department it is assigned.

Service Management also checks in COMPLAINTS whether the complaint has occurred before. If so, it is decided whether the solution is offered immediately to the complainant. There are two possibilities:

- If the solution is satisfactory for the complainant, this is recorded in COMPLAINTS and the complaint can be closed.

(continued)

- If the solution is not satisfactory for the complainant, the complaint is registered with the department that should solve the problem. This will not be Service Management, but a product department. There are several different types of product departments. For example, there is one responsible for all bus processes, another for all metro processes, and a third for all travel products.

The product department accepts the complaint and a complaint handler is allocated responsibility for resolving the complaint. Acceptance takes place by sending an e-mail to Service Management. If the responsibility is not accepted, the Head of Service Management is informed and the complaint is referred to a more senior manager.

The complaint handler analyzes the complaint, using the information obtained, and proposes a solution. The analysis results are recorded in COMPLAINTS.

The solution resulting from the analysis is tested before it is implemented. This reduces the chance that the solution has undesirable side effect, or may not even work at all. It may occur that the solution is so simple that testing it is not necessary. Such judgment is made by Service Management.

If the test does not show the desired results, the complaint handler re-analyzes the complaint and uses the result to find a new solution. All test results are recorded in COMPLAINTS.

The solution is then implemented in the real world and recorded in COMPLAINTS. It is reported back to the complainant by the complaint coordinator, including a request for approval. At that point there are two possibilities:

- The solution has actually solved the complainant's problem: the complaint is closed in COMPLAINTS.
- The solution did not resolve the complainant's problem: the complaint coordinator retrieves new information from the complainant and sends it to the product department for renewed analysis.

Upon acceptance of the solution by the complainant, the complaint coordinator determines the end time. The lead time is the difference between the end time and the start time specified in the complaint notification. If there is no response from the complainant, the complaint coordinator determines whether the complaint should be considered solved. The complainant's response is recorded in COMPLAINTS.

The complaint is closed in COMPLAINTS. All available information regarding the solution is stored in the system for future analysis, for example by examining trends (which can then be further analyzed) or weaknesses in processes and information systems.

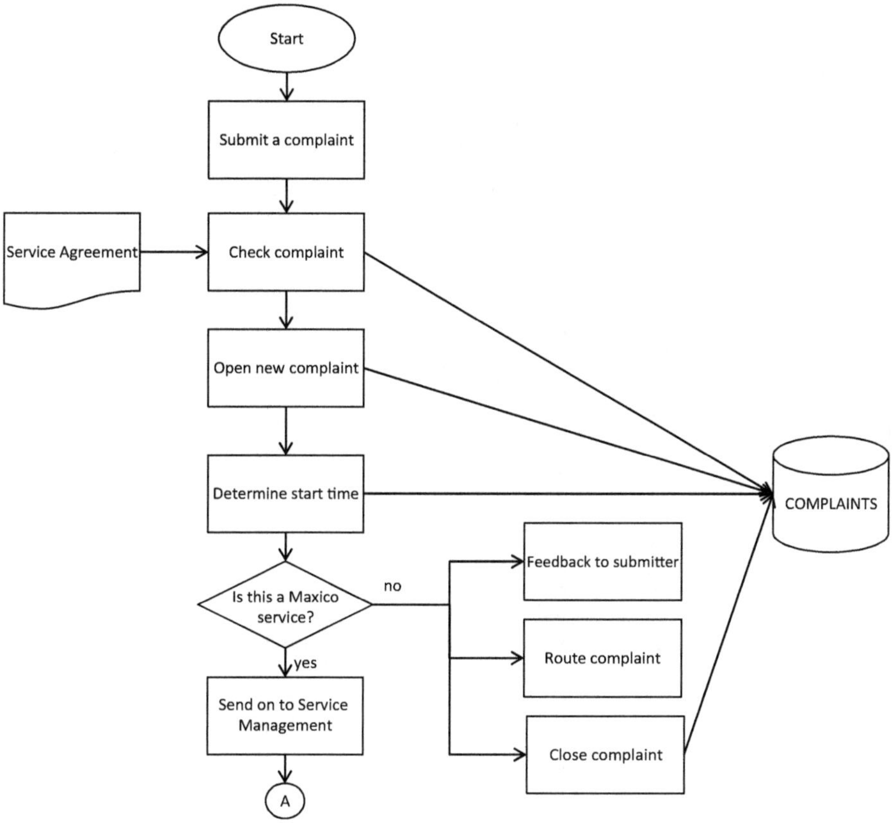

Fig. 4.1 Submitting a complaint

If we would want to depict this process in a flowchart, one process designer would outline the workflow differently compared to another. One may describe many details, another few; one may leave out a particular activity, while another will include it. And so on. In this example I opted for just one process modeling approach, leaving it open for discussion whether this is the right way.

In Fig. 4.1, the complaint handling process begins with the receipt of a complaint.

In Fig. 4.2, Service Management classifies the complaint and determines the priority.

In Fig. 4.3, the complaint handler takes on the case.

In Fig. 4.4, the complaint is resolved.

And in Fig. 4.5, the solution of the complaint is reported back to the complainant.

In Sect. 4.3 we will examine this example in more detail. But before we do that, we must first become acquainted with the different cooperation levels in an organization.

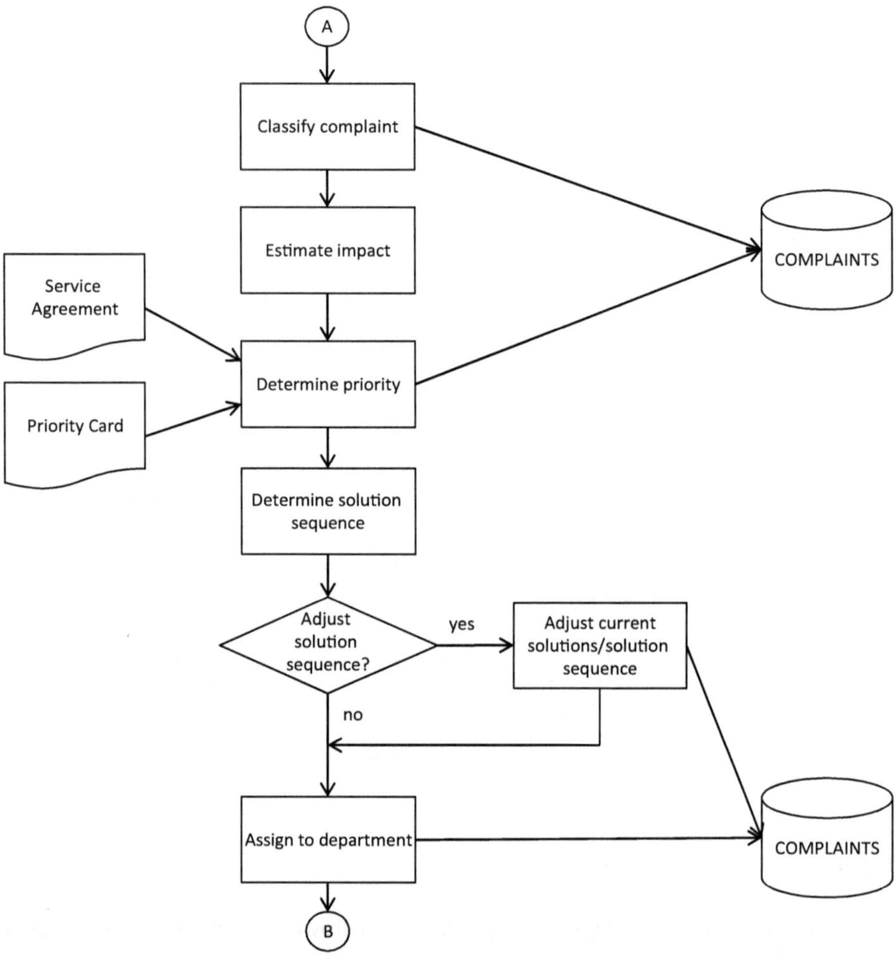

Fig. 4.2 Classifying and prioritizing a complaint

4.2 Cooperation Levels in Organizations

Cooperation is what makes an organization an organization. Such cooperation is achieved through communication. Using written or spoken language, people make agreements. If we want to thoroughly investigate cooperation, we need to know more about how people communicate and about the information that is shared when communicating.

In organizations, there is a huge amount of communication going on, but not all of it is relevant to achieving business goals. In order to avoid drowning in non-relevant communication, it is important to distinguish between different levels. The founders of Enterprise Engineering have proved that communication and

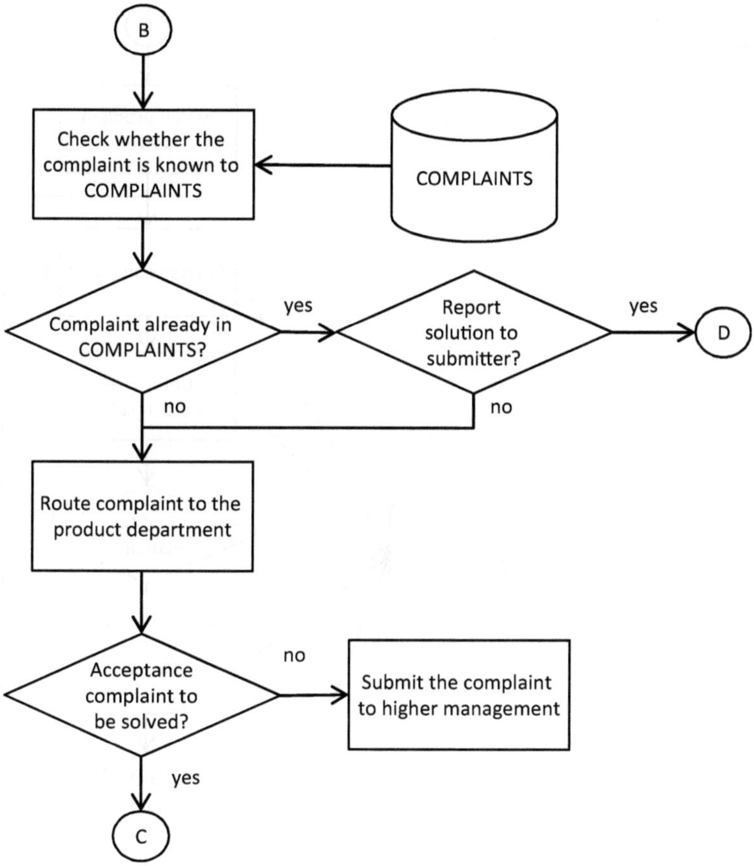

Fig. 4.3 Routing a complaint

information take place at different levels. We need to make a clear distinction between

a) the 'datalogical',
b) 'infological', and
c) 'ontological' levels.

Based on the distinction between these levels, we can determine what is fundamental for organizational cooperation—and what is not. Anything of fundamental importance for the organization is called the *essence*. Everything else is derived from that *essence*. As we will see later, the infological and datalogical levels are not essential, but the ontological level is.

I'm afraid it is impossible to avoid using the terms datalogy, infology and ontology from the Enterprise Engineering discipline's strict set of definitions. But

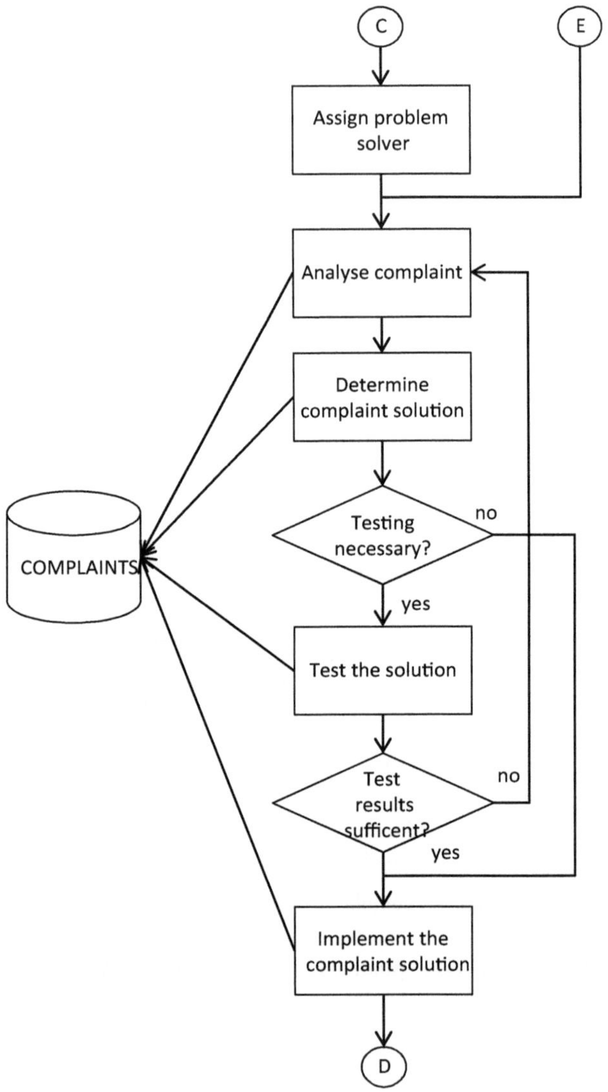

Fig. 4.4 Resolving a complaint

don't worry, I will explain these difficult terms. In Fig. 4.6, we see the three levels in organizations in relation to each other. The deepest level is the ontological.

4.2.1 The Datalogical Level

The datalogical level relates to the format in which information is captured: it is about the carrier of information, rather than the information itself. When we look at e.g. function and process descriptions, reports, minutes, e-mails, oral messages and

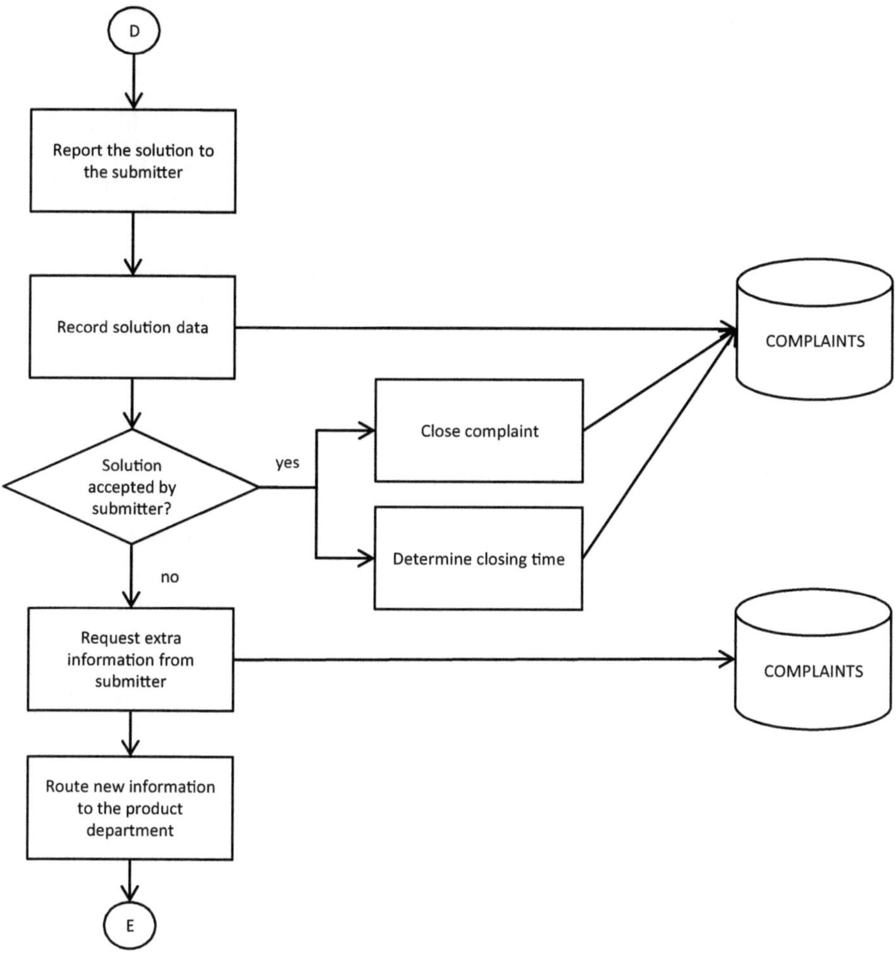

Fig. 4.5 Reporting and closing a complaint

customer databases, the datalogical level is about the format of the information. Is it written on paper? Is it captured electronically? Is the information exchanged orally? And so on. When an Enterprise Engineer refers to the datalogical level, he does not mean the information itself, but the shape in which it is molded. Copying a report, typing a process description, printing a chapter of this book, destroying confidential information are all datalogical activities. These kinds of production are referred to as *datalogical production*. The information carriers are used in communication. They play a role in making agreements, but they do not bring them about. They only support communication.

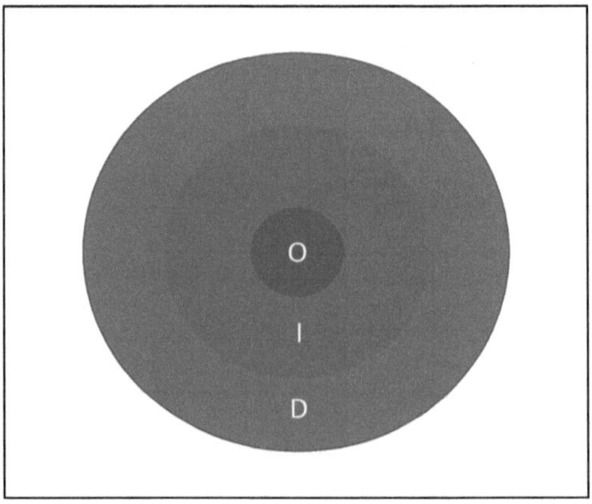

Fig. 4.6 The three levels in organizations

4.2.2 The Infological Level

The infological level looks at the information itself. The distinction from the datalogical level is that the former refers to the content, not the format. It is perhaps somewhat abstract to separate information itself from the format in which it is presented. Of course, you cannot convey any information without an information carrier (datalogy). But it is important to make this distinction because the decision to store information on information carrier A or carrier B should be taken only after agreement on the content of the information itself. Confidential information is not published on a public website; in a restaurant I would not expect to get a menu on a USB-stick. Hence, the information content largely determines the format in which it is presented.

A second characteristic of the infological level is the fact that it refers to information derivable from existing data. Consider the contents of a company's annual report. Or the calculation of the balance sheet total, if all posts and the business rules for calculation are agreed.

We call the production of information *infological production*. It was Börje Langefors who, in his book *Theoretical Analysis of Information Systems* [1] from the late 1960s, first introduced the concept of 'infology'.

4.2.3 The Ontological Level

If we look at an organization at a datalogical and infological level, we see that something is missing. Of course, with the division between the infological and datalogical levels, we take into account that information is used and created in an organization (the infological level), and that this information also becomes shaped in some way (the datalogical level). But what we are missing, for example, is making the decision to create information. Or giving the order to develop a new product. 'Making decisions' and 'giving orders' are not classifiable in the infological or datalogical levels. So where do they belong?

People who cooperate make agreements with each other and commit to them. In fact, people in organizations 'produce' agreements. We call this *ontological production*. Only by making agreements can people get things done. If a baker wants to bake bread or cakes, he will have to make agreements with others (or, if he is autonomous, agreements with himself) to purchase and maintain the ovens. He must make arrangements to buy ingredients. In order to sell the bread, he makes an agreement with the customer that the latter will pay after he has received the bread. And the bread itself must be baked.

The infological and datalogical levels follow the ontological level. They are important, but without ontological production—an agreement—nothing happens. In order to be able to bring about an agreement, it is usually convenient to have access to information. For example, the baker who places an order with the wholesaler benefits greatly from a list of ingredients (infology), which may be available on paper (datalogy). But without bringing about an agreement to place an order, the ingredients list is useless, and so is the piece of paper it is printed on—in this case, the most important thing is that the baker commits to making an order. What he orders exactly, and in what format he books his order with the wholesaler, is subordinate to the agreement itself. Hence the ontological layer forms the essence of organizations. The word 'ontology' is well chosen in this case. It means 'the being', 'the essence'.

Since agreements make an organization what it is—effective cooperation between people—bringing about agreements belongs to the essential, or ontological, level.

Infological and datalogical production are all derivable from earlier decisions. Ontological production, however, is original and not derivable from any other fact. Thus, the decision to buy a loaf of bread does not come from a list of ingredients but is mainly motivated by the need to provide our bodies with food.

By plotting the organization's ontological level, the Enterprise Engineer tracks all the practical agreements that are being made including their interrelationships. We will see that all agreements are made in the same universal way, always and everywhere. In other words, coming to an agreement always follows one and the same pattern, which we will hereafter call a 'transaction'. All these arrangements, together with their patterns and interrelationships, constitute the organization's *construction*.

4.3 Levels in Maxico

In the previous section I described the distinction between the ontological, infological and datalogical levels. In this section we will apply these three levels to an example of the Maxico transport company. One way to do this is by means of emphases in the text and in the flowcharts. I will highlight ontological aspects by bold, infological aspects by bold-italic and datalogical aspects by underline emphases. Using the process description already described for complaint handling, we get the below result. Note that relevant aspects are emphasised only once, other occurrences are not.

Complaints about the transport service are **filed** by e-mail or phone by a 'complainant'. A complainant can be one of the following parties:

- A customer.
- An internal department of Maxico.
- A third party (such as the National Railway Company).

After receipt, the complaint coordinator makes a first scan to **check** whether the complaint relates to the services of Maxico. The complaint coordinator is the one that follows a complaint from start to finish and ensures that everyone who plays a role in the complaint handling process completes their responsibilities in a timely manner. Both in case of an internal department and a third party, a so-called ***service agreement*** is consulted regarding the applicable complaint handling **agreements**. These may be, for example, agreements related to the complaint handling lead time.

For each complaint received, the complaint coordinator ***opens*** a new complaint in the COMPLAINTS information system, and enters the ***basic facts of the complaint***. If the complaint is not the responsibility of Maxico, this will be **fed back** to the complainant and the complaint will be **forwarded** to the appropriate party. That information is also recorded in COMPLAINTS, and the complaint is regarded as **closed**.

If the complaint needs to be resolved by Maxico, based on the information obtained, the complaint is ***classified*** by the Service Management department. The classification includes the ***complaint type***. For example, does the complaint relate to service operation, such as a bus that is late, or to bad/undelivered customer service, such as an incorrect travel advice?

Subsequently, Service Management **determines** the complaint priority, based on the service agreement, the **estimated impact** and ***urgency*** indicated by the complainant. This is done using a formula provided on a Priority Card.

The complaints solution order is ***determined*** based on the priorities. With each new complaint received, it is **assessed** whether a change in the solution order is needed, and whether ongoing activities need to be temporarily set aside in order to give precedence to solutions with a higher priority.

It is Service Management's responsibility to record the complaint as completely as possible in COMPLAINTS, at the same time noting to which department it is **assigned**.

(continued)

Service Management also ***checks*** in COMPLAINTS whether the complaint has occurred before. If so, it is **decided** whether the solution is offered immediately to the complainant. There are two possibilities:

- If the solution is satisfactory for the complainant, this is recorded in COMPLAINTS and the complaint can be closed.
- If the solution is not satisfactory for the complainant, the complaint is **registered** with the department that should **solve the problem**. This will not be Service Management, but a product department. There are several different types of product departments. For example, there is one responsible for all bus processes, another for all metro processes, and a third for all travel products.

The product department **accepts** the complaint and a complaint handler is **allocated responsibility for resolving the complaint**. Acceptance takes place by sending an e-mail to Service Management. If the responsibility is not accepted, the Head of Service Management is informed and the complaint **is referred** to a more senior manager.

The complaint handler ***analyzes*** the complaint, using the information obtained, and **proposes a solution**. The ***analysis results*** are recorded in COMPLAINTS.

The solution resulting from the analysis is **tested** before it is implemented. This reduces the chance that the solution has undesirable side effects, or may not even work at all. It may occur that the solution is so simple that testing it is not necessary. Such **judgment** is made by Service Management.

If the test does not show the desired results, the complaint handler re-analyzes the complaint and uses the result to find a new solution. All test results are recorded in COMPLAINTS.

The solution is then **implemented** in the real world and recorded in COMPLAINTS. It is **reported back** to the complainant by the complaint coordinator, including a request for **approval**. At that point there are two possibilities:

- The solution has actually solved the complainant's problem: the complaint is closed in COMPLAINTS.
- The solution did not resolve the complainant's problem: the complaint coordinator retrieves new ***information*** from the complainant and sends it to the product department for **renewed analysis**.

Upon acceptance of the solution by the complainant, the complaint coordinator determines the ***end time***. The ***lead time*** is the difference between the end time and the ***start time*** specified in the complaint notification. If there is

(continued)

no response from the complainant, the complaint coordinator **determines** whether the **complaint should be considered solved. The complainant's response** is recorded in COMPLAINTS.

The complaint is closed in COMPLAINTS. All available ***information regarding the solution*** is stored in the system for future analysis, for example by ***examining trends*** (which can then be further analyzed) or weaknesses in processes and information systems.

Both the red bold text and the flowcharts indicate that the essence forms a very small part of all the information recorded about Maxico. This allows us to focus on what is really important in an organization or process flow.

In Fig. 4.7, the complaint handling process begins with the receipt of a complaint.

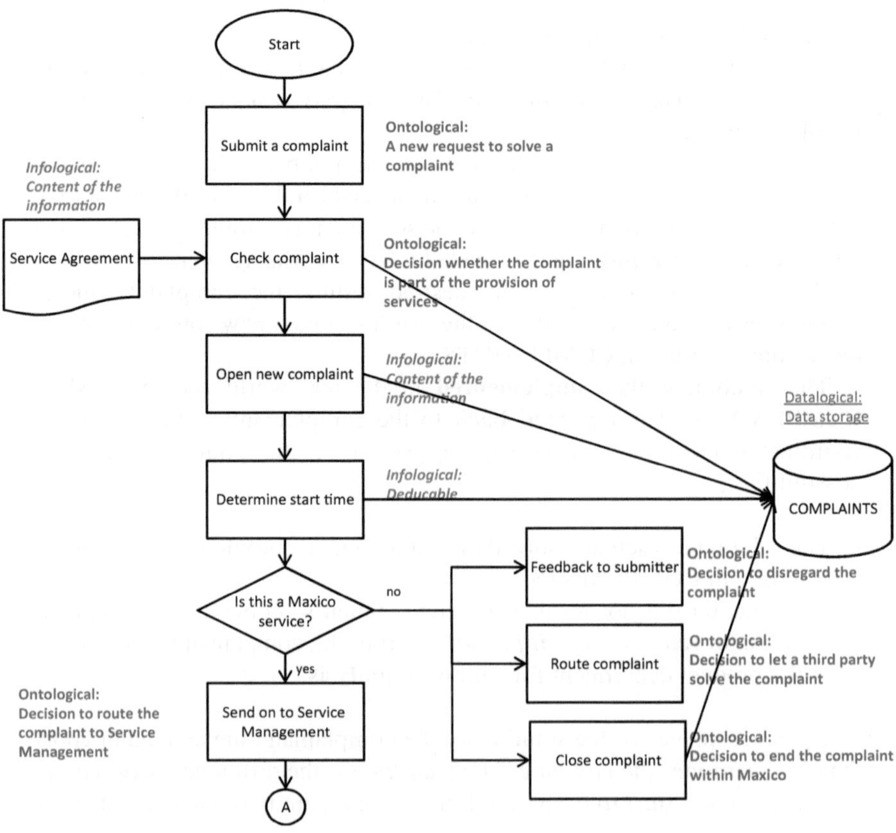

Fig. 4.7 Submitting a complaint—split in the three levels

In Fig. 4.8, Service Management classifies the complaint and determines the priority level.

In Fig. 4.9, the complaint handler takes on the case.

In Fig. 4.10 the complaint is resolved.

And in Fig. 4.11, the solution of the complaint is reported back to the complainant.

The text and the process flows demonstrate how little essence remains if you only consider the ontological aspects.

We have now begun to explore the ontological elements in an organization's process, in this case Maxico's complaint handling process. However more work is needed in order to expose the complete construction of an organization. We will see that it is important for the construction to understand 'which employees in which functions' are involved in the ontological activities. We will introduce new concepts in order to better define employees and functions: 'transactions' (the above introduced concept, which in Enterprise Engineering is used to define the way in which agreements are brought about in organizations) and 'actor roles' (the roles which are fulfilled by the 'actors' or 'acting persons', active in the organization). But first I will introduce a number of models, used to expose the ontological level.

4.4 Modeling the Deepest Level

As mentioned, the ontological level in an organization (or an organizational unit, depending on the chosen area of investigation) is made up of 'transactions', i.e. the way in which agreements in organizations are brought about, and 'actor roles', i.e. the people acting in an organization. Once all these transactions and actor roles are thoroughly defined, the Enterprise Engineer has a huge amount of data at his disposal. There are four models with which he can depict the cooperation (consisting of transactions) between different actor roles (persons acting in an organization): the Construction Model, the Process Model, the Action Model and the Fact Model. Together, these four models present a complete picture of the entire set of agreements (the ontological level) which constitute an organization. Together, the Construction Model, Process Model, Action Model and Fact Model form the 'essence' of an organization. The Construction Model, also called the 'construction', forms the core of the essence. Once this construction has been found, the organization is like a 'white box': the construction of an organization is exposed just as the construction of a car is visible once the hood is opened.

Of the four models, the Construction Model and the Process Model are the most important. The other two models (the Action Model and the Fact Model) primarily serve to check the completeness and the accuracy of the Construction Model. Moreover, the Action Model and the Fact Model are rather technical—hence I leave them outside the realms of this book. For further reading, I recommend *Enterprise Ontology* by Jan Dietz, which explains clearly why these four models are used to display the full working reality of an organization. Granted, it may seem impossible to fully expose an organization, but it is indeed possible for the *formal*

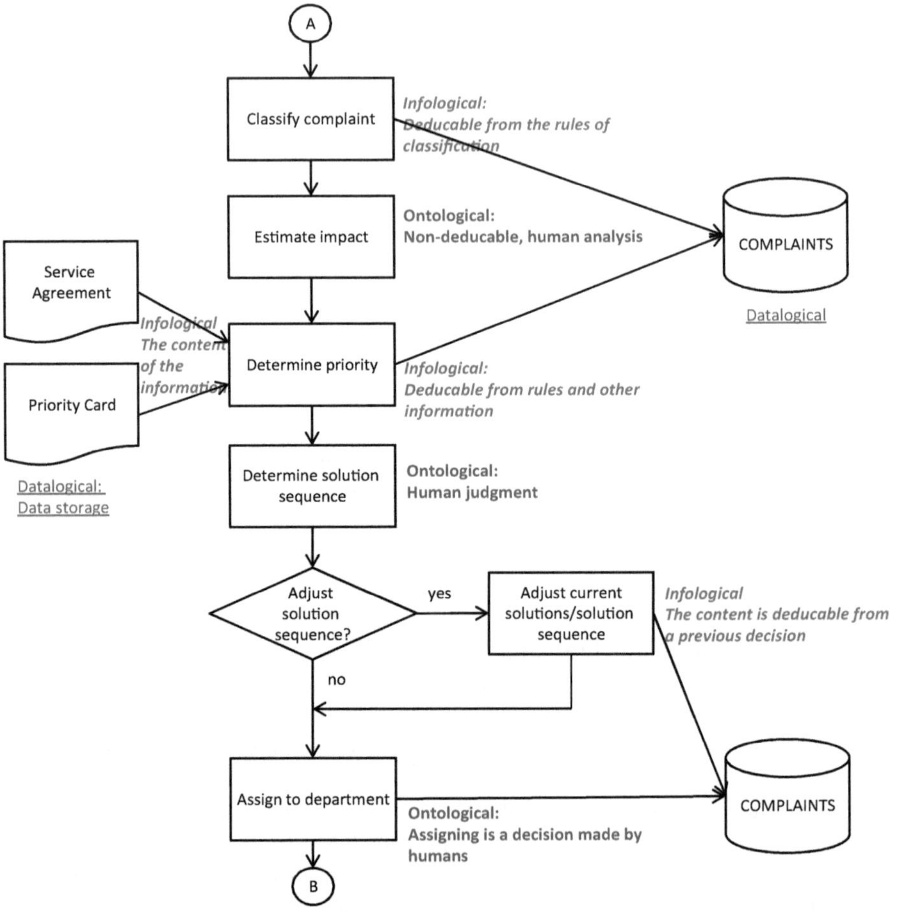

Fig. 4.8 Classifying and prioritizing a complaint—split in the three levels

organization. By *formal* organization I mean all the operations within an organization which are aimed at the goals of this organization. The *informal* organization, the domain of the (individual) interests and other socio-technical interactions in organizations, can also be made transparent with the help of actor roles and transactions, but this would hardly make sense because these types of transactions are so volatile; interests can change quickly. See Sect. 8.4 for more background on the informal organization. Ergo, the entire formal organization can be modeled with the help of Enterprise Engineering.

In the Construction Model, the actor roles, transactions and the environment of an organization are displayed. Firstly, a table (among others) is prepared in which all transactions are recorded, numbered T01, T02, et cetera. For example: T01 'new customer registration'. The result of the transaction is also numbered and described. For example: R01 'new customer registered'.

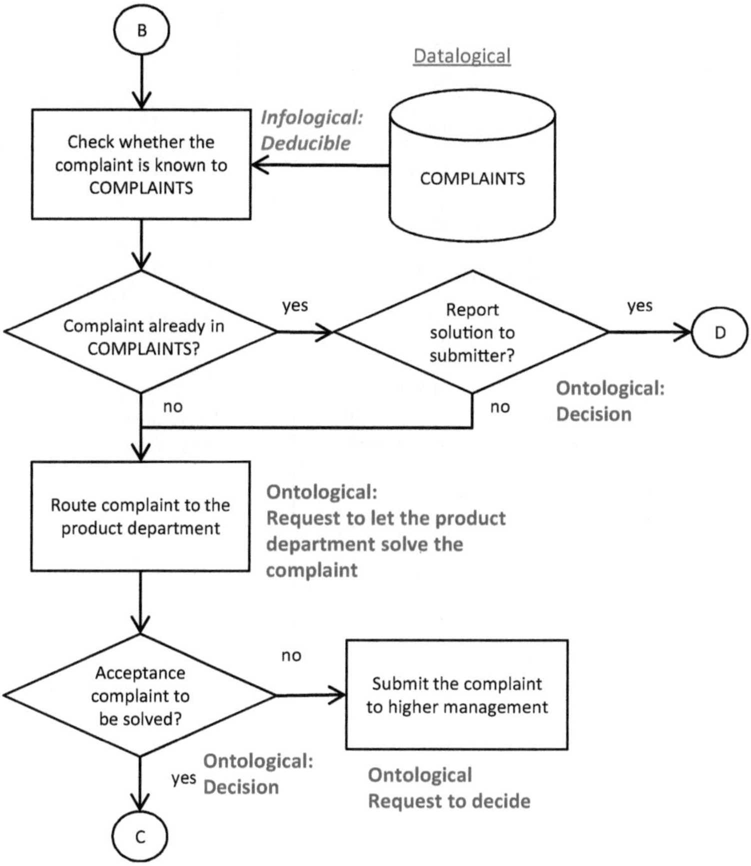

Fig. 4.9 Routing a complaint—split in the three levels

The Construction Model is then based on this table. The Construction Model shows all transactions that take place, how they relate to each other, which actor roles are involved (and thus interact with each other), and the environment in which the transactions are initiated. A department or an outside party, such as a supplier, is an example of such an environment. In Sect. 4.5 we discuss the Construction Model further.

The Process Model, as the name implies, indicates the (business) processes in an organization, in the form of a series of transactions. The Process Model places the transactions in a specific order (in contrast to the Construction Model, which does not order transactions). Since the Process Model represents a sequence of successive steps, it is a process. The Process Model logically follows the Construction Model and therefore there is nothing superfluous nor anything omitted that should

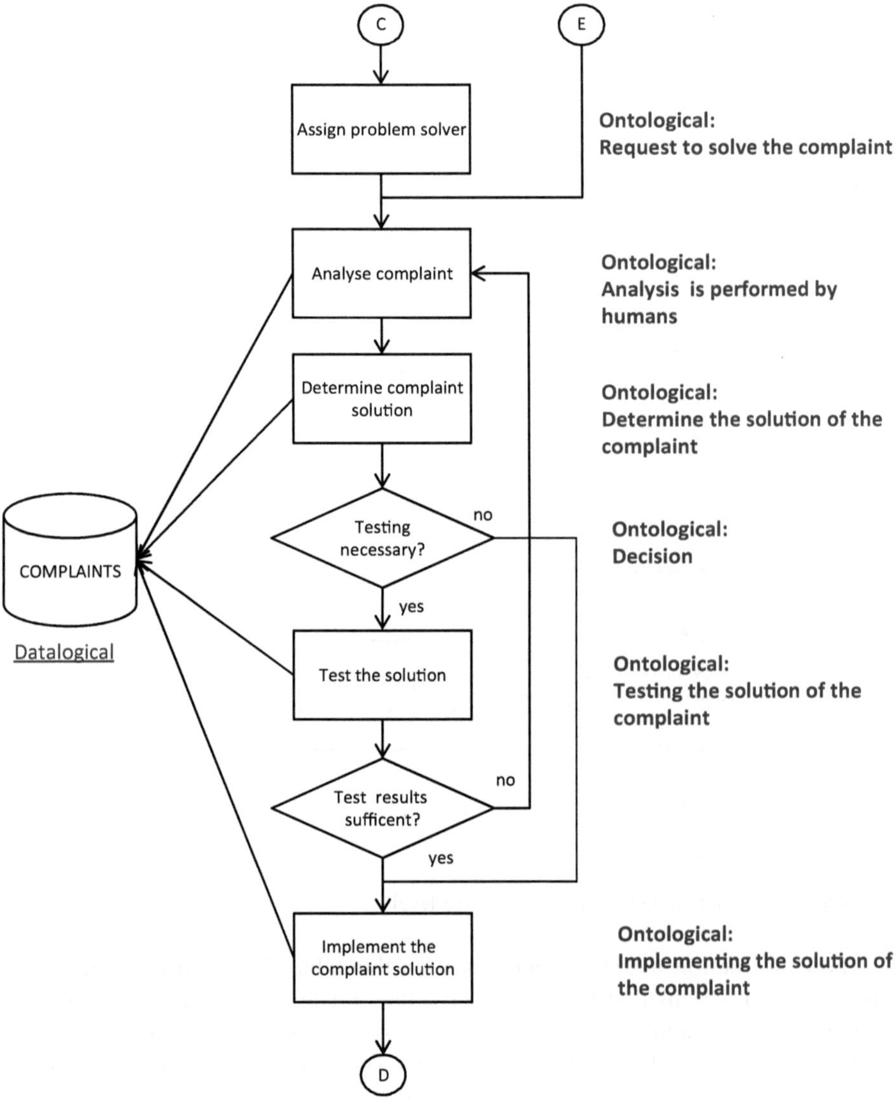

Fig. 4.10 Resolving a complaint—split in the three levels

have been included. In other words, the processes described in the Process Model are the essential processes. The Process Model provides process designers with a means to decide objectively and exactly what a process diagram must contain.

In Fig. 4.12 we see the four different models of the ontological level in an organization, in relation to each other.

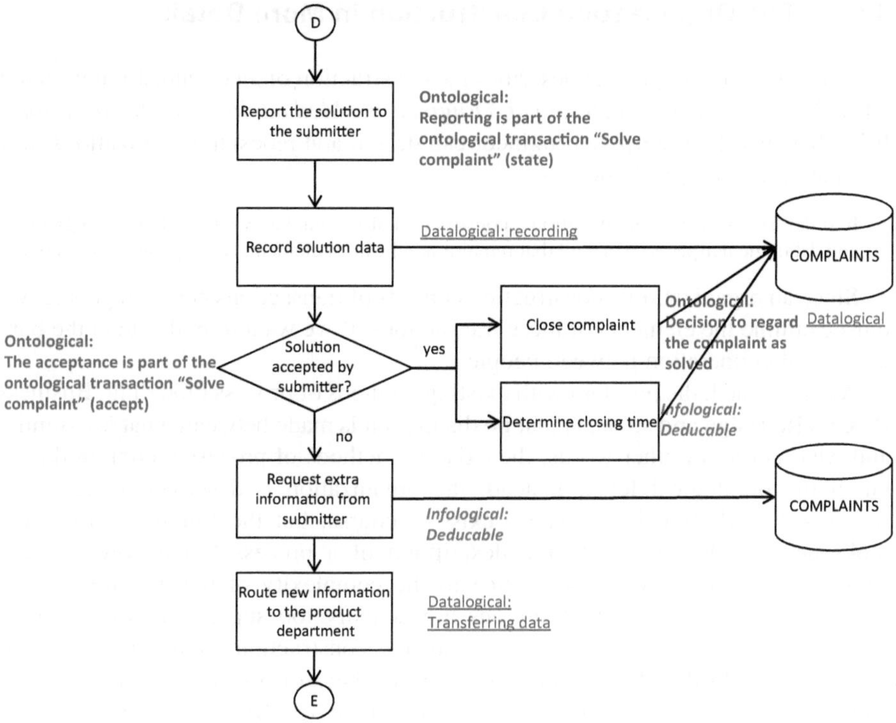

Fig. 4.11 Reporting and closing a complaint—split in the three levels

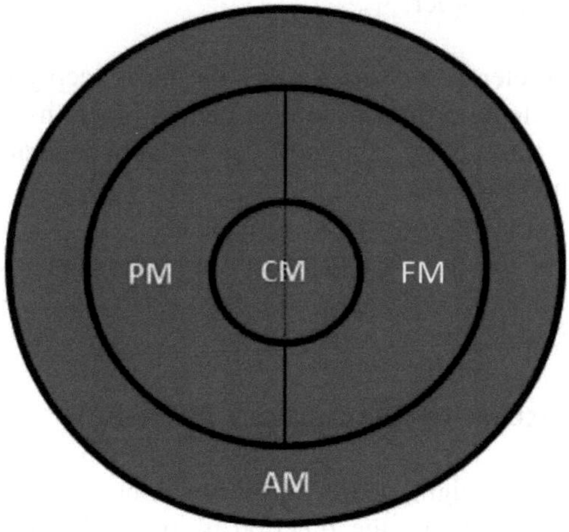

Fig. 4.12 The four distinguished models of the ontological level

4.5 The Organization Construction in More Detail

The modeling technique that describes the construction of an organization is called DEMO©. This stands for *Design and Engineering Methodology for Organizations.* It implies that the Enterprise Engineer can design and redesign organizations, just as an architect designs buildings.

> In order to describe an organization's construction, it is completely stripped of infological and datalogical aspects so that all that remains are the most elementary cooperation aspects.

Since an organization's construction consists of transactions between people, we must examine the elements of these transactions, if we want to understand the bare essence of cooperation between people.

As mentioned, the problem with existing methods of process modeling, Business Process Redesign and so on, is that no distinction is made between what is essential and what is not. In other words, the existing methods of process modeling do not target the ontological level. Instead, the organization is described in terms of 'processes' and 'functions'. These exist of course, but the Enterprise Engineer looks deeper. He realizes that a description of a process 'on its own' is too simplistic; which does not do justice to the complexity of the operation of an organization. Processes and functions are actually too superficial because they *follow* the construction—but they *are* not the construction itself. Processes and functions should therefore be analyzed further, because we do not know exactly what is going on within those functions and processes. By further dissecting the processes and functions, the engineer discovers the construction of the deeper-lying ontological level.

In this section I will discuss the most important concepts of Enterprise Engineering, in the following order:

- From function to 'elementary actor role': the most deeply laid responsibilities, authorities, competencies that are hidden behind functions.
- From process to 'transaction': the most deeply laid cooperation agreements that are hidden behind processes.
- Cancellation patterns in cooperation agreements.
- Other elements of a construction: organizational boundaries and information banks.

4.5.1 From Function to 'Elementary Actor Role'

In an organization, at first sight, people fulfill functions. A function has a name, and somewhere in a docket or file, the function is 'precisely' described. If someone wants to know what employee A does within an organization, he can consult that file, as is often claimed. Of course this does not in any way reflect what employee A actually experiences within the recurring complexity of the organization, in which all kinds of different colleagues and possibly external, sub-contracted employees interact in all sorts of different situations. These job descriptions are a simplification of reality. They are useful, because they provide some degree of definition, but it is possible to describe a function much more clearly. Looking at the roles that individuals play provides a much more complete and nuanced picture. When cooperation is carefully examined, it appears that the actors, or acting people, assume different roles in the organization. This is a phenomenon that occurs in every organization.

To take a simple example, what role does a baker take on when he sells bread to a customer? In that case, he is a 'bread seller'. However, before he can sell bread, he has to bake it first. He is literally a 'baker'. Before he can bake the bread, he will have to buy the ingredients. Then he is the 'ingredient purchaser'. And so on. Thus the 'baker' function continually assumes different roles. The 'baker' function is an example of a 'composite actor role' because it consists of several 'elementary actor roles'. An 'elementary actor role' is a single role: it produces only one ontological fact. In the example of the baker, an elementary actor role would be the role of 'bread seller' (only sells bread and nothing else), or that of 'baker' (only bakes bread and nothing else). You might think that an elementary actor role such as 'baker' can be further broken down into even smaller actor roles.

Actually this is not the case, but to gain a deeper understanding of this, more insight is needed into the aforementioned Action Model, the State Model, and the relationships between the four models of the ontological level in an organization. The substantiation of this is reflected in the aforementioned book by Professor J. Dietz (*Enterprise Ontology*).

Hence, a function like that of 'baker' says relatively little about the separate roles fulfilled by the 'baker'. Only when the composite actor roles of the 'baker' are accurately considered, can we see what the baker is doing in detail. This allows us to the track all of the different 'cooperations' that the person in question actually deals with.

Every organization appears to be a complex set of composite and elementary actor roles, which cooperate through a specific pattern.

In the terminology of Enterprise Engineering, these patterns of cooperation and collaborative partnerships are called 'transactions'. An example this would be the ordering of bread or the ordering of ingredients. I will expand on this later. For now, it is important to look at what is required to fulfill an elementary actor role adequately. If one of the following three components of the elementary actor role is missing, this will have consequences for the proper execution of a transaction:

- Responsibilities—An employee who takes on an elementary actor role, thereby takes the responsibility to perform the actions necessary for a particular transaction. In other words, the employee who performs an elementary actor role, has been given the responsibility by the organization to deliver the corresponding transaction through his actions, and he must also be willing to meet this responsibility. *For example*: the baker must and wants to sell the bread.

- Authorities—Whoever occupies an elementary actor role must also have a mandate empowering him to execute a transaction. *For example:* the baker must have the authority from his boss to sell the bread.
- Competencies—Finally, the relevant skills must be in place for the implementation of an elementary actor role. In other words, does the employee have the skills needed to function well in his respective elementary actor role? Some conditions, such as tools and time, also come under the 'competencies' heading. For example, the baker must be physically able to sell the bread (not lying sick in his bed), and the baker must have time to sell the bread (rather than being busy baking the bread while the customers are waiting in the shop).

If even one of these three components of an elementary actor role—responsibility, authority and competence—is lacking, the organization, in this case the bakery, has a problem. An incompetent salesperson may charge too much or too little; an under authorized salesperson is not allowed to make bulk deliveries to wholesale clients. A salesperson who does not fulfill his responsibilities, for example, sells too little or nothing.

Responsibility, authority and competence are intrinsic components to the elementary actor role. A role can be performed (competencies), may be performed (authority) and must be performed (responsibility). If one of these elements is missing, the product cannot be delivered. The word 'elementary' in the phrase 'elementary actor role' refers to the minimum possible combination of responsibilities, authorities and competencies necessary to deliver the product. If a 'bread salesman' has slightly less responsibility, authority or competence than is required to sell bread, he will be unable to fulfill the actor role. The elementary actor role is responsible for delivering one and only one product type and has precisely enough responsibilities, authorities and competencies to deliver this product type. In Fig. 4.13 we see how an actor role is represented.

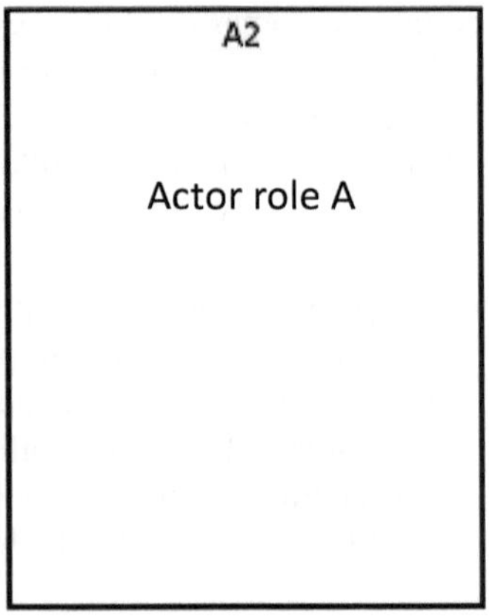

Fig. 4.13 The representation of an actor role

4.5.2 From Process to Transaction

As mentioned previously, Enterprise Engineering calls an organization's cooperation method a 'transaction.' A transaction is not a process. Just like a job description, a process description is a rather broad generalization. In fact a process is not elementary, but divisible: a process consists of a series of related transactions. In any organization, these transactions take place continuously and they always follow the same pattern. The transaction is the universal way in which people reach agreements [2]:

1. Role A (an actor in his role, such as 'customer') starts the transaction with a *request*.
2. Another, role B (e.g.; the baker who sells bread in his elementary actor role as 'bread seller') agrees to fulfill that *request* and makes a *promise*.
3. Next, the production takes place (for example, a loaf of bread is packaged in a bag).
4. After this, B *states* the product to be ready and offers it to A (the baker offers the bread to the customer).
5. A completes the transaction by *accepting* this offer.

Activities 1, 2, 4 and 5 bring about the cooperation: they are called *coordination activities*. Together with the product delivery (activity 3), this forms the so-called *standard transaction pattern*. There are many ways of performing these coordination activities: implicitly, explicitly, indirectly, directly, ambiguously and so on. Sometimes one can get the impression that these four coordination activities (*request*, *promise*, *state*, *accept*) are not all performed in a given transaction, but that is an illusion.

With a single nod or by reaching for his wallet, a customer indicates to the baker that he has *accepted* the bread. All actions that play a role in a transaction can be performed implicitly—except for the *request*. When the customer enters the bakery and just looks at the baker, the baker will not know what is required of him. In principle, all other activities can be performed silently. Since humans are social beings, we understand what is happening from the context and from the non-verbal communication. Yet it is often more comfortable to present the actions explicitly. A friendly 'here you are, Sir' when offering the bread is quite different than if the transaction is completed in silence.

Any process that takes place in an organization can be interpreted in the described manner: each process consists of (a) several transactions, each of which further consists of (b) coordinating activities carried out by (c) two or more actors in specific elementary actor roles, and (d) the relationship between these transactions. Eventually there is a result, which leads to production. By 'production' I mean a combination of activities called a *production act*, whereby something new, a product, is created. There are two types of products:

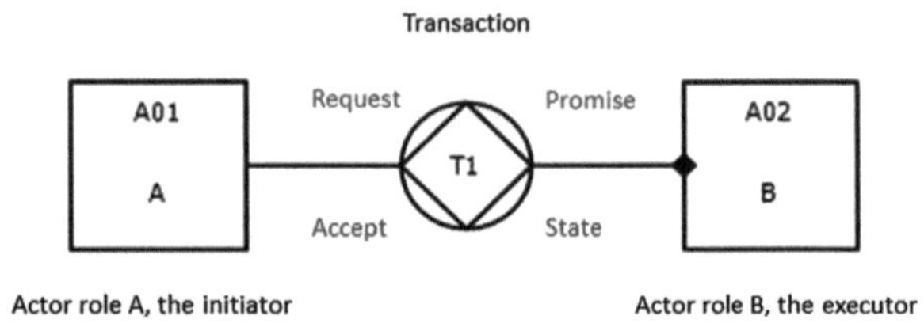

Fig. 4.14 Two actor roles and their binding transaction

- A physical product such as a sliced loaf of bread.
- A non-physical product such as a decision or a judgment.

For the sake of simplicity, I only deal with the first type of production in this book.

Figure 4.14 shows an actor role (A) that initiates the transaction by making a *request*. Actor role B is the one that delivers the product (the black dot). The transaction is as follows: A makes a *request*, B makes a *promise*, B provides the *product*, B performs the *state* (i.e. states the product to be ready, handing it over to the customer) and A *accepts* the offer. Thus a transaction consists of three phases. The order phase (*request* and *promise*), the execution phase (the product is created) and the result phase (*state* and *accept*). The person who makes a *request* is the initiator of the transaction. The person who makes a *promise* is the executor.

The unique feature of displaying this cooperation in the form of a separate transaction, unlike, for example, flowcharts in which processes are displayed, is that it presents a much more complete picture. It indicates:

- Which actor role starts the transaction
- The name of that actor role
- The name of the transaction
- The actor role which is tasked with providing the product
- The name of that actor role

Naturally, within an organization, there are countless interrelated transactions. There are several models available to provide a clear oversight of this complex system. Later I will discuss the models that the Enterprise Engineer has at his disposal.

4.5.3 Cancellation Patterns

Transactions can also be canceled according to a fixed pattern. A *request* may be refused, for example: in such a case, the executor or (intended) executor of the transaction decides not to complete the *request*. Another form of cancellation is when the executor does respond to the *request*, but sees his *state* rejected because the initiator (the person who made the *request*) does not agree with what is offered to him (again this may be for various reasons, such as poor quality, high price, late delivery).

Initiator: makes *request*
Executor: *promises* or *refuses*
Executor: performs the *state*
Initiator: *accepts* or *rejects*

All four of the coordination actions can be cancelled. On cancellation, the transaction no longer follows a normal course. There is no longer a standard transaction pattern, but a *cancellation pattern*. There are four cancellation patterns or four ways in which a transaction may be canceled.

1. The *request* may be canceled—for example, a customer of the bakery has requested some bread, but withdraws the *request* because on closer inspection, he wants another loaf of bread.
2. The *promise* may be canceled—the baker discovered, after the *promise* was made to the customer, that the requested bread is no longer available.
3. The *state* may be canceled—for example, the baker thinks that the bread he has offered is not at all suitable for the customer (perhaps because it is not fresh), and he withdraws the *state*.
4. The *accept* may be canceled—the customer realizes that the bread does not meet his expectations, and revokes his decision to accept it.

In case of cancellation, the transaction is completed, but without a result. However that does not matter; this is part of mutually bringing about agreements between people. No one will be surprised should the baker say that there is no more bread left and that, therefore, no bread can be delivered/sold. The standard transaction pattern, combined with all cancellation patterns, forms a universal cooperation pattern. Whenever people bring about transactions with each other, it is done in this way. It never happens in any other way, or even in a different order, and no new types of activities are performed other than the activities in this universal pattern. This pattern is valid now and it will be a thousand years from now. Sociologists have uncovered this pattern, and as long as we humans remain social beings, we continue to stick to this exact pattern. As do the people who work in organizations.

Because there are so many possible ways of progressing through a transaction pattern, it has been decided (for the sake of clarity) to use one fixed symbol for transactions in the Construction Model. This symbol is the transaction symbol: a circle containing a diamond (see Fig. 4.14)—in this symbol, the diamond represents the product, the circle the four coordination activities of a transaction.

4.5.4 Other Construction Elements

There are two more elements that can be displayed in a construction: organization boundaries and information banks.

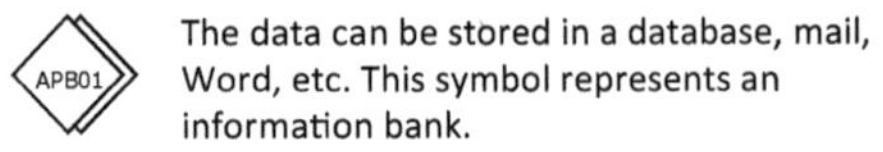

Fig. 4.15 The representation of an information bank

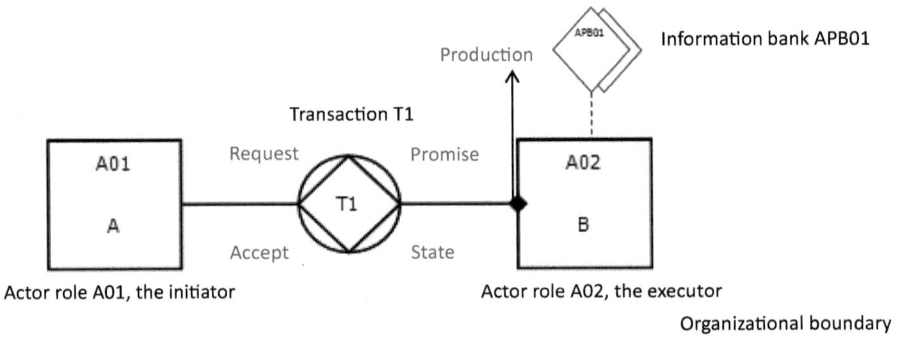

Fig. 4.16 All elements of a construction model

An organization boundary limits the Enterprise Engineer's area of investigation. It may consist of the entire organization, or a division, or several sub-divisions. It is up to the Enterprise Engineer to determine where the organization border can best be drawn. This has to do with the underlying problem that is to be solved. The organization boundary comes from the systems theory: it is the boundary of the system under consideration.

An information bank refers to information that is necessary to enable the carrying out of a transaction. The transaction symbol itself (the diamond in the circle) includes information exchange, but this only implies information that is available to the actor roles themselves. The information bank, on the other hand, refers to information that should remain available after the transaction has ended, or existing information that should be accessible during a transaction.

For example, the bakery salesperson records the number of loaves sold in a spreadsheet; every day, he writes in this spreadsheet how many loaves were sold that day. The ingredients purchaser (in this example, this role is fulfilled by someone other than the salesperson) needs this information in order to purchase the right amount of ingredients. The spreadsheet containing the information about the bread sales is then recorded in the Construction Model as an information bank. In Fig. 4.15 we see how an information bank is displayed.

And finally in Fig. 4.16, all the elements of the Construction Model are illustrated in one picture.

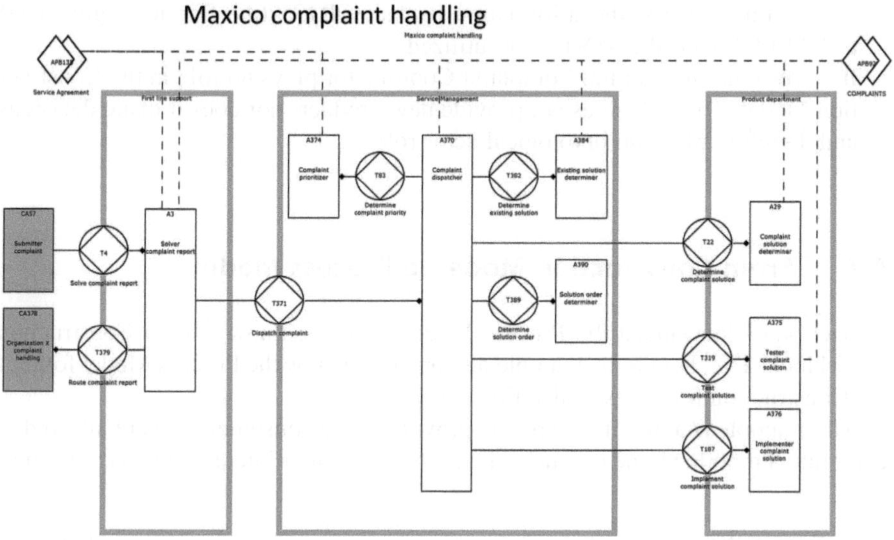

Fig. 4.17 The Maxico complaint handling construction model

4.6 The Construction Model of Maxico

Now that we know how to put a Construction Model together, we can look at one in practice, namely the Maxico Construction Model, see Fig. 4.17.

In the Maxico Construction Model, we see the following elements:

- There are three Organization parts: First line, Service Management and the product department.
- The complainant registers his complaint with the First line (T4).
- If the complaint cannot be resolved by Maxico, it is redirected to another organization (T379).
- If Maxico is indeed deemed able to handle the complaint, the complaint is picked up by Service Management (T371).
- The so-called Complaint Handler determines the priority of the complaint (T83), determines whether there is already an existing solution (T382) and determines the resolution order of the complaints (T389), if applicable.
- If there is an existing solution to the complaint, then the First line is informed of this (T371) and in turn the First line keeps the complainant up to date regarding the solution (T4). In both cases this forms the transaction *state*.
- If there is no existing solution, the product department takes over the complaint from Service Management and determines the complaint resolution (T22), tests the complaint resolution (T319) and implements the complaint resolution (T187). The complaint is subsequently closed by successively making the *states* of T187, T371 and T4.

- During this process, the information banks APB131 (the Service Agreement) and APB93 (COMPLAINTS) are utilized.
- It is worth noting that the Complaint Coordinator plays no role in the Construction Model. This role does not provide new products nor does it make decisions and therefore plays no ontological actor role.

4.7 From Construction Model to Process Model

In Sect. 4.4 I introduced the Process Model. Now we know what a Construction Model looks like, here is an example demonstrating how the Process Model follows the Construction Model (see also Fig. 4.18).

The example is a mobile telephony provider. As a consumer, you are allowed to enter into a mobile phone contract (transaction 1 = enter into a contract) as soon as

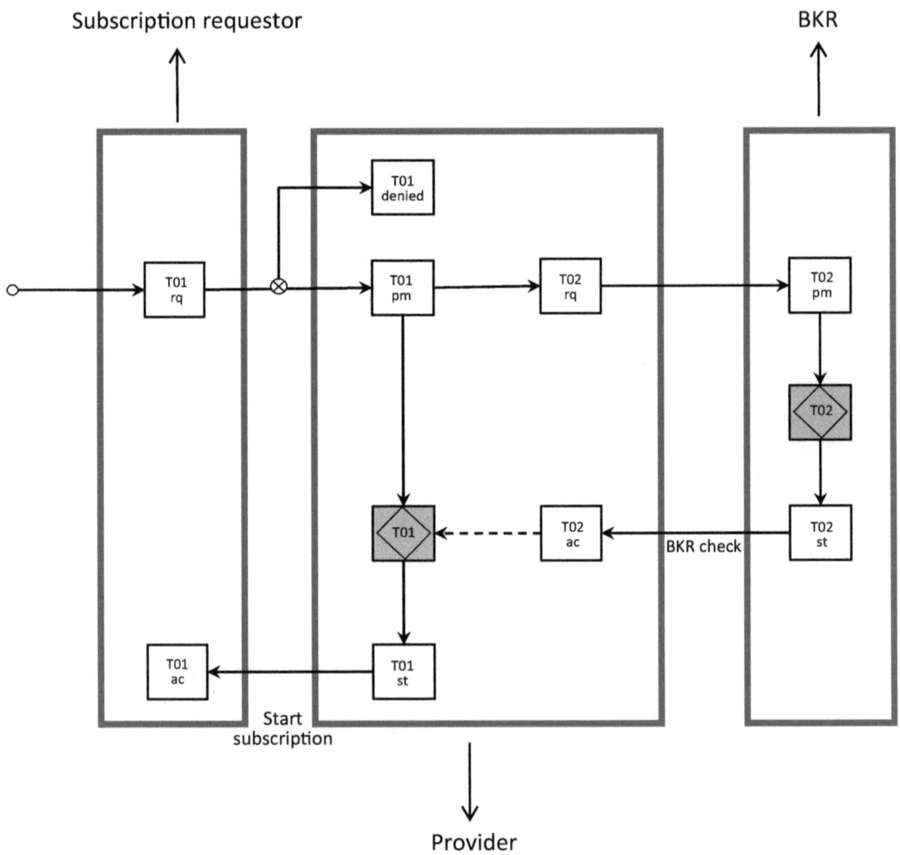

Fig. 4.18 The process model is derived from the construction model

the provider has checked your credit status with the Bureau of Credit (transaction 2 = BoC check).

In the Process Model, the sequence of the two transactions is as follows:

- *Transaction 1*—A *request* to enter into a contract; *promise* to take the *request* into consideration. It is possible that an internal audit of the social security number of the applicant shows that this person is on this provider's 'black list', for example because he has not paid his bills in the past. Then the *request* is immediately lapsed and the transaction is stopped (T01 in Fig. 4.18). If the service provider's internal audit of the potential new customer does not set off any alarm bells, then transaction 2 is put in motion: BoC check.
- *Transaction 2*—A *request* is sent from the provider to the BoC for a so-called BoC check. This *request* is responded to (by BoC) with the *promise* that the check will be performed. The BoC check is delivered by the BoC itself, after which it is followed by the *state*: BoC offers the provider the result of the executed BoC check. This transaction is completed with the *accept* of the result by the provider. If the BoC check did not provide any objectionable information regarding the potential new customer, the provider can now complete transaction 1: *Transaction 1*—Contract creation in the system, contract *state* and finally the contract *accept*.

The condition—that a BoC check with a positive result should be received before a contract can be completed—is not visible in the Construction Model, but it is in the Process Model.

4.8 Focus Areas of the Enterprise Engineer

Since an organization may be very large and a specific problem may not manifest itself everywhere within an organization, it is important to introduce a focus when Enterprise Engineering. In this section, I will describe two choices regarding the areas of consideration.

4.8.1 AS IS and TO BE Models

It is always recommended to start by exposing the current or AS IS situation. This concerns the Construction Model of an organization, or usually a part of the organization, meaning a representation of the organization construction *at that particular moment*. Based on the AS IS situation, the Enterprise Engineer can start redesigning the cooperation. The actual areas to be improved depend on the objectives and the correction feasibility of the flaws identified. The end result is a Construction Model of the TO BE situation, which can then be implemented.

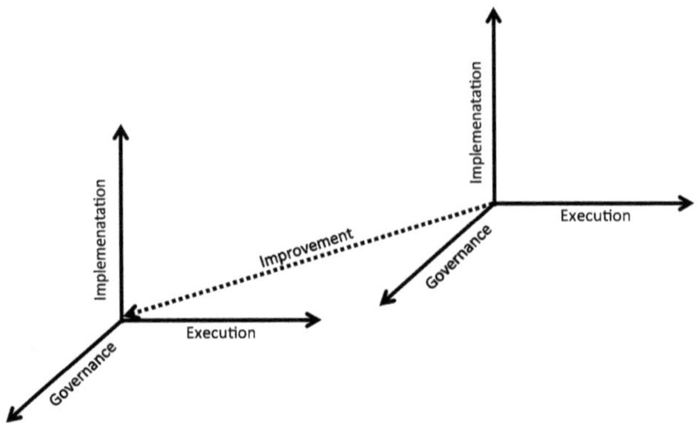

Fig. 4.19 The four dimensions in an organization

4.8.2 Organization Dimensions

An Enterprise Engineer will rarely, if ever, model a complete organization construction. Organizations are too large for this and customers are often too cautious. During exploratory discussions with the stakeholders, such as the management, they usually identify a specific department, unit or field in which problems need to be solved. The Enterprise Engineer then determines his scope.

However, at that point the scope is not yet complete. Once it is determined which department(s) will be subject to analysis, it is necessary to determine which dimensions of that area are of interest. An organization can be viewed in four dimensions and, in each, dimension-specific transactions take place (see Fig. 4.19):

- *Governance*—This is the dimension in which management decisions are made. This mainly concerns the strategic and tactical level in organizations.
- *Implementation*—Divisions, departments, programs and projects must be implemented after they are defined, for example, when an organization has just started or is expanding. This leads to temporary, specific elementary actor roles and transactions. They belong to the implementation dimension.
- *Execution*—This is the world of the employees performing their jobs. It refers to daily organizational operations.
- *Improvement*—All transactions, relating to actor attempts to autonomously implement organizational improvements, fall within the improvements dimension.

Governance, implementation, execution and improvement are called 'dimensions' because together, they fully describe the organization. There is no conceivable area in an organization that does not fall under one of these four dimensions. Moreover, the dimensions do not show any overlap: for example, an actor role in the implementation dimension will not appear in the governance dimension and vice versa.

4.9 The Strategy as a Starting Point for the Organization Construction

What has the construction of an organization to do with its strategy? And how do vision, mission and policy relate to this 'cooperation architecture'? Vision, mission, policy and strategy do not follow the construction, but they precede it.

Before considering precisely how the vision, mission, policy and strategy are associated with the construction, we must first decide what we mean by vision, mission, policy and strategy. They are indeed common concepts, but in almost every management book you will find a (slightly) different interpretation.

A *vision* is a picture of the future that you want to create. The vision not only outlines how the future should look, but also how to actively contribute to achieving it. Out of the vision comes the *mission*. By mission I mean: what the organization—inspired by the vision—will stand for in the future. In order to go from the current situation to the desired situation (mission), concrete *goals* are formulated. Thirdly, the *policy* must ensure that the way of working strictly adheres to these goals. Since there are countless conceivable ways of working towards the goals, management must impose strict organizational frameworks. These frameworks, which together make up the policy, are meant to avoid proliferation in the way of working. They set out the path that the employees should follow. Finally, the *strategy* is *the way* in which the organization wants to achieve its goals, within the policy frameworks of course.

An example of vision, mission, policy, strategy and goals:

Vision	Before the end of the 1960s there will be a man on the Moon.
Mission	That man will be an American.
Policy	NASA will build the rocket which will take the man to the Moon.
Strategy	First experiment with animals in space, and if it appears to be working, we send people into space. Only then do we set foot on the Moon.
Goals	The milestones on the way to the mission, such as 'a dog in space', 'a monkey in space', 'a man in space' and 'a man on the Moon'. Goals are specific markers in the strategy.

In Fig. 4.20 the concepts of vision, mission, policy, strategy and goals are visualized.

Every organization starts out with ideas about vision, mission and strategy. These *precede* the form chosen for cooperation—the organization construction—in which the strategy must be put into practice. This construction is only built after the formulation of ideas, if there is no organization yet. But mostly, organizations already exist. And strategies often evolve. The question that is rarely asked, but certainly should be asked, is: can the organization effectively handle the chosen strategy? Should anything change in the organization construction? And if so, what? And are there certain things we have to bring to an end, due to a particular strategy?

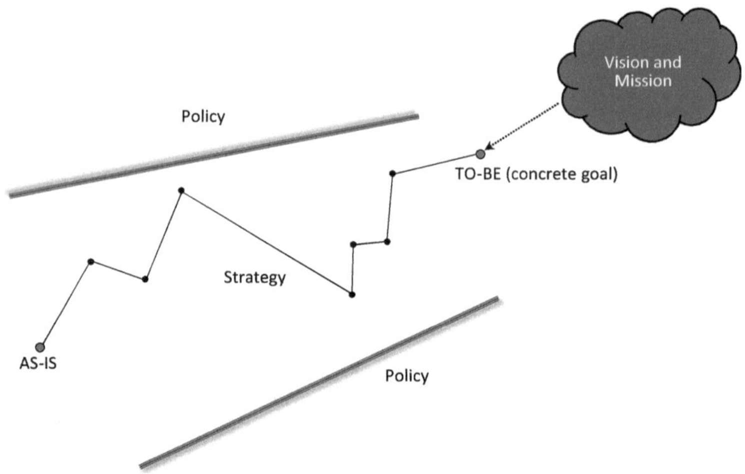

Fig. 4.20 The concepts of vision, mission, policy, strategy and goals

All too often, wonderful strategies are developed, without any awareness of whether their implementation in the organization is feasible. The result is that very little of the strategy ends up being implemented. A program is started, a lot of money is put into it, maybe some goals are even achieved, but in the long run you will see no return on investment.

For example, in an attempt to cut costs, an organization decides to decentralize a number of tasks that are currently performed centrally. Certain departments now have to perform these tasks themselves. In the beginning, this succeeds, but after a few years these departments appear to have developed totally different ways of working that do not align. Subsequently, the organization suffers greatly: changes that need to be rolled out across various departments take a long time, fail and/or cost a lot of money. It may very well be that, eventually, the decentralized performance is more expensive than the previous central implementation. The organization may then decide to revert to the tasks being performed centrally again—and then in a subsequent cost-cutting exercise, to decentralize them again. This centralization and decentralization cycle occurs in all organizations and takes about 10 years on average. Consultants earn a lot of money from this: if an organization is centrally arranged, they invariably recommend ‘decentralizing’. And for decentralized organization, the advice is to ‘centralize’. A report in which that advice is expressed does not come cheap. And the process of centralization-decentralization can easily cost millions. What a waste of capital.

With the help of a Construction Model, the existing organization construction can be exposed. Based on that model, adjustments can be made to the construction to facilitate the strategy realization.

Thus a new Construction Model is created. It is crucial, when designing this new Construction Model, to discontinue all activities that do not contribute to the strategy realization.

For example, an organization decides to develop an outsourcing strategy (perhaps because they foresee a knowledge gap within a few years). In this case, the construction of the organization is extended with at least *one* transaction: outsourcing (of service and/or product). Often there will be many more transactions, because in outsourcing, for example, you must also perform many more checks to determine whether the party who is sub-contracted really understands your requirements. At the same time, by making this choice for outsourcing, a number of transactions are skipped: transactions related to the internal delivery of the product or service which now is procured externally. In this way, the organization construction is adapted to the strategy.

4.10 The Organization Construction as Starting Point for Organizational Design

It is important to have a thorough organization Construction Model, because the construction determines the nature of its processes, and not the other way around.

We have seen that *processes* are derivable from the construction. The construction of a car determines the process of getting in: if the steering wheel is on the left, as the driver, you have to get in on the left. That also means that you keep the key in your right hand, open the door with your left hand and put your right leg in first. Try this when the steering wheel is on the right hand side.

Opening the door will still work, but putting your right leg in first? The construction, in this case the construction of the car with the steering wheel on the left, determines the entry process.

Secondly, the *structure* of an organization follows from its construction. By structure, I mean the functional division into departments, units and so on. This is called an *organization chart.* If it turns out that there are a huge number of transactions between two or more parts of the organization, one can ask whether it is not much more convenient to combine these departments. The reverse is also true: if one intends to split up a department, it is a good idea to take its transactions into consideration. Thus it can be determined (a) whether it is actually useful for that department to be split up (or not), and (b) if it is decided to split it up, where the best 'cut' can be made. If the cooperation is challenged by the proposed split, one should think again. A striking example is the splitting up of the Dutch Railways (responsible for the trains, scheduling, travelers, et cetera) and ProRail (responsible for the infrastructure on which the trains can travel), which turned to be quite a poor decision. Thus, the Construction Model not only prescribes the processes, but also the ideal organization chart.

Thirdly, the organization's management can define the *governance structure* from the construction. If, based on the construction, new departments are to be created, they must also be managed and controlled. In addition, there are transactions on the organization boundary, or on department boundaries, which by definition require additional management attention. Based on the construction, it can be determined which transaction clusters must be managed by *one* manager, for example because they are logically related. Hence, the construction gives the board an objective tool with which to determine how the organization can best be structured. In a reorganization, for example, there may be many different ways of redrawing the organization chart. But which hierarchical structure is the best?

Fourthly, the construction can also clarify *external relationships*—the transactions that lie on the boundary of the organization—in order to improve operational cooperation with external parties. Without a Construction Model, the complexity of such transactions is often underestimated, because it is assumed that there are only two actor roles in place: 'customer' and 'supplier' respectively. The common approach is that a customer and a supplier make mutual agreements, often drawn up in a *Service Level Agreement* (SLA). By doing so, organizations often expect to have a complete and clear set of agreements. But this view of the 'customer and supplier making an agreement' lacks many fundamental actor roles and transactions. Who monitors and controls the supplier performance?

Who delivers the results for each service? These are probably all different elementary actor roles, and each one may be inadequate for some reason. Such problems, however, never come to light when the relevant organizations (customer and supplier) consider the cooperation as only one transaction. This explains why outsourcing and SLAs lead to so many problems in practice; parties base their ways of working on an oversimplified representation of reality.

Fifthly, the Construction Model can help to define which processes in the organization are candidates for automation. An organization usually benefits greatly from an optimum alignment of automation and cooperation processes. Since a Construction Model provides a good cooperation oversight, use of the model can help determine which transactions require ICT support and in which

Fig. 4.21 The construction model of a private club

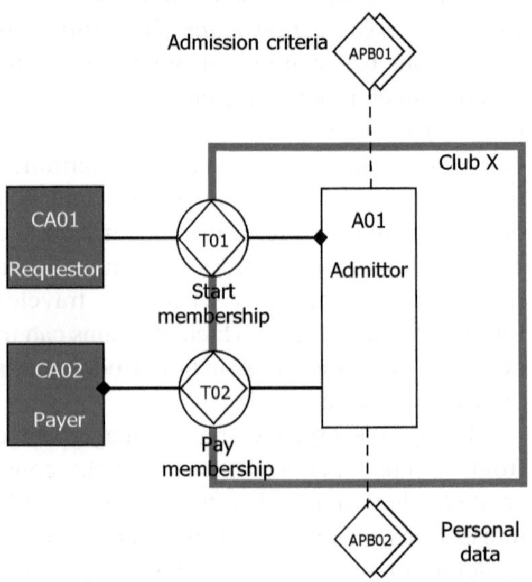

form (e.g., an application), and which do not. And information banks could be considered to be implemented in the form of ICT solutions. In Fig. 4.21 we see, by way of example, a Construction Model of a private club. Anyone can apply to become a member of the club (CA01), but he/she is only accepted as a member if all admission criteria (APB01) are met. Once the candidate member meets these criteria, he/she becomes a member and pays membership (T02). In order to make someone a member, personal data is needed, which is recorded in an information bank (APB02).

The admission criteria should be clear to everyone, also for people outside Club X; in this way, the basis on which decisions are made regarding who is allowed to join, is clear to everyone. In other words, the construction determines the information required, here in the form of information bank APB01. This information bank could be automated, for example, as a PDF list on a website.

This transparency does not apply to the club members' personal data. This data falls within the boundary of Club X and must not be accessible to people outside the organization. This data is, for example, stored and maintained in a database and kept confidential.

Finally, the Construction Model forms the basis for the function descriptions of the actor roles, and therefore the organizational functions (the sum of all actor roles). For example, Fig. 4.21 shows that the person at Club X who admits the new members has the following responsibilities, authorities and competencies, among others:

- Responsibilities:
  - Check whether a candidate member is eligible to be a member
  - Collection of the membership fees
- Authorities:
  - Decide whether someone is allowed to join (not accountable to others)
  - Access the personal data of (new) members
  - Access bank account details of Club X
- Competencies:
  - Can apply admission criteria in practice
  - Can record personal information in a database
  - Shows integrity
  - Is meticulous

References

1. Langefors, B. (1973). *Theoretical analysis of information systems*. Auerbach: Studentlitteratur.
2. Franke, R. H., & Kaul, J. D. (1978). The Hawthorne experiments: First statistical interpretation. *American Sociological Review, 43*, 623–643.

Enterprise Engineering in Practice

5

In the foregoing four chapters we have seen that organizations, at their most elementary level, consist of actor roles and transactions that are connected by way of a universal pattern. We refer to the total pattern as the construction of the organization. The construction describes the internal operation of an organization. The discipline that analyzes organizations at elementary level is called Enterprise Engineering. In this chapter we will look into how Enterprise Engineering is applied in practice. In practice, there appear to be 22 types of error patterns of cooperation. I will discuss each of them. I will then explain how Enterprise Engineering can be applied, i.e. the *Way of Working*.

5.1 Finding Flaws in Organizations

The detailed AS IS Construction Models uncover the construction of the organization at a glance. One of the major advantages of this overview of the organization is that the models reveal the construction flaws. The major part of the problems faced by organizations is caused by these flaws, which are basic errors in the universal pattern of cooperation. We refer to them as Elementary Construction Flaws (ECFs). The word 'flaw' refers to a deviation from the universal pattern of cooperation. The word 'construction' indicates that the flaw is visible in the Construction Model. And the addition 'elementary' indicates that these flaws take place at the smallest possible level; they are not caused by an even deeper underlying problem. They are caused by elementary actor roles and the transactions between them and therefore form the true root cause of the problems.

At first glance ECFs seem to be caused by a lack of initiative, lack of knowledge, conflicting characters et cetera but this is not always the case. By nature, people are willing to cooperate. The flaws are embedded in the underlying organizational construction, in the working patterns. Human responses to these types of errors are the result of the error patterns rather than their cause. They complicate the cooperation even more.

T. Janssen, *Enterprise Engineering*, Management for Professionals,
DOI 10.1007/978-3-319-24172-2_5

Just resolving all the ECFs in an organization will greatly benefit an organization, particularly in terms of efficiency and effectiveness, but it will also improve the atmosphere at work. I have drawn up an overview of all 22 ECFs that occur in practice. You will see how easily you can detect and restore them thanks to the analysis in terms of elementary actor roles and transactions.

Just to clarify: these 22 ECFs all occur in practice (my personal inventory). You could refer to them as archetypes, but such is already explicitly incorporated in the qualification 'elementary'. These 22 ECFs do not overlap, they are all different. There are probably no additional ECFs, but there is no evidence to support such. As such, this list provides an adequate tool for detecting the root causes of problems in organizations.

I have drawn up a list of all 22 ECFs as I think it will enable you better to detect ECFs in your own organization. For readers who are first learning about ECFs, it may be slightly boring to read all the texts about ECFs in one go. It might be better for these readers to read the ECFs bit by bit and alternate them with the other parts of this chapter.

5.1.1 ECFs with Regard to the Role of the Initiator

5.1.1.1 ECF 1: Missing Initiator ('Redundant Work')

There are people who perform work nobody asks for anymore. The work, i.e. the transaction and the result, was necessary in the past, but the work has become redundant due to a reorganization, a change program or a new manager who places the focus elsewhere. Still the older version of the job description still lists it as one of the activities that needs to be done, or people continue doing it out of habit.

For example: when I introduced a new Requirements Management Tool (RMT) in the organization where I worked at the time, the management wanted to track the use of this tool. This would allow them to monitor and adjust it if necessary. After a while the RMT was fully implemented, but people still kept on reporting about it, even though the management did not read those reports anymore. In other words: the management (the initiator of this transaction) did de facto no longer make the *request* to perform the transaction of 'reporting on the RMT'. But because it was never *formally* agreed that the reports were no longer required, people continued to submit reports: this turned out to be unnecessary, as the management did not need these reports anymore. Figure 5.1 depicts the way in which ECF 1 is represented.

5.1.1.2 ECF 2: Unclear Initiator ('Unclear for Whom')

Some process documents hold various functions responsible for starting a process. Even if this responsibility has not been documented datalogically, i.e. if it not

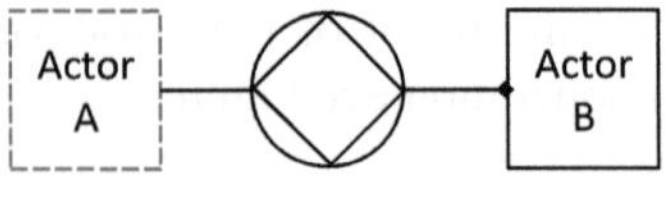

Fig. 5.1 ECF 1: Missing initiator

explicitly stated in the process description/manual, practice shows that some functions feel responsible for the start. However, this multiple responsibility perception does not automatically mean that all those different actor roles will exactly initiate the transaction that starts the process. The lack of clarity can even lead to none of the actors initiating the transaction, which will result in the process not being initiated at all.

For example: Someone working at a bakery needs to check once every 3 days whether new orders need to be placed with the supplier. It is usually the owner who issues the *request* to do this, but sometimes it is issued by his wife, and regularly by an employee who has been working for the bakery for 15 years. One Saturday there are hardly any pies in stock, and it's the day before Mother's day. A new employee had not checked the stock and therefore had not ordered pies. The owner and his wife had assumed that the permanent employee would have asked the new employee to do this, but the permanent employee thought that the owner or his wife would do so, as he himself was busy giving instructions to a temporary employee. Figure 5.2 depicts the way in which ECF 2 is represented.

5.1.1.3 ECF 3: Missing Responsibility Initiator ('Not Interested in the Result')

Sometimes the initiator does not feel responsible for initiating the work, as a result of which the request (and/or the acceptance) is carried out hastily, is incomplete or is of a bad quality. The results are obvious. This problem can be caused by delegating the task to an incorrect role.

For example: the team leader is given an additional task: collecting information on budget expenditure for a (bi-monthly) report. However, he thinks this job is awful and boring. He does not take the time to really dive into the matter and collect the right information or assess it properly. The result is an incomplete report that will negatively affect the budget allocated for the following financial year. Figure 5.3 depicts the way in which ECF 3 is represented.

5.1.1.4 ECF 4: Missing Authority Initiator ('Unauthorized Work')

Each actor role must have the authority (i.e. mandate) to perform his work. This not only applies to the executing actor role, but also to the initiating actor role. After all, someone placing a request without being authorized to do so runs a great risk that the executing actor role will refuse to comply with the request (because that

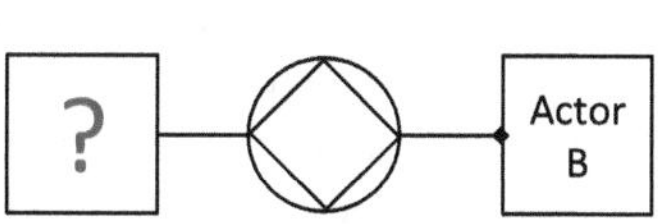

Fig. 5.2 ECF 2: Unclear initiator

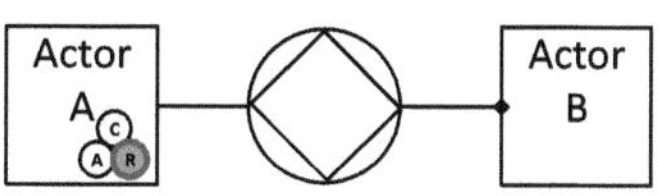

Fig. 5.3 ECF 3: Missing responsibility initiator

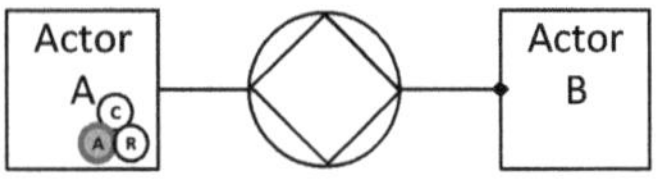

Fig. 5.4 ECF 4: Missing authority initiator

executing actor role does not want to run the risk to do something which later turns out not to be permitted).

For example: if the actor role wants to take 50 € from the cash machine, but lost the pin code, the machine will certainly refuse to complete the transaction and will not submit cash. The machine will assume that he is not authorized to take money out of the account. Figure 5.4 depicts the way in which ECF 4 is represented.

5.1.1.5 ECF 5: Missing Competence Initiator ('Don't Know What You Want')

If the initiating actor role does have the authority and responsibility, but lacks the competence (due to a lack of education, skills, experience et cetera.), the transaction cannot be completed, or will be of an insufficient quality level. The lack of competence can relate to the formulation of the right request, but also to the acceptance of the result (is the actor capable of reviewing the result critically?).

For example: a manager has an audit performed based on a renowned quality model. However, he has no experience with these types of audits, and the audit turns out not to answer his questions. He has asked the wrong audit questions or maybe even hired the wrong bureau. Figure 5.5 depicts the way in which ECF 5 is represented.

5.1.1.6 ECF 6: Split Request and Accept Initiator ('Requestor and Acceptor Have Incompatible Views')

An actor role that initiates a transaction with a request is usually also responsible for the acceptance. After all, a person making a request usually wants to know whether a request has been complied with; only then will he accept that what is offered.

Sometimes the act of 'acceptance' is delegated. This can be useful. But when the person to whom the act is delegated is unaware of what the original request entailed, it is possible that he accepts the result—which could have negative consequences should it become clear that the result does not suffice. It is also possible that actor role A does not know exactly what to ask for because B is always responsible for the acceptance and is therefore better aware of the details.

For example: A man and a woman are in a clothing shop. The man (actor role A) needs a new sweater and asks his wife to purchase it because she is fashion conscious. Actor role A therefore initiates the transaction. The woman complies with his request. In the clothing shop an employee (actor role B) conducts the promise and the state. The wife who has been delegated the coordination act of the acceptance, forgets that her husband is allergic to wool. She buys a beautiful sweater, but gives her husband blood-shot eyes. Figure 5.6 depicts the way in which ECF 6 is represented.

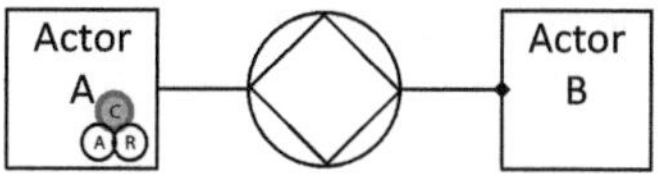

Fig. 5.5 ECF 5: Missing competence initiator

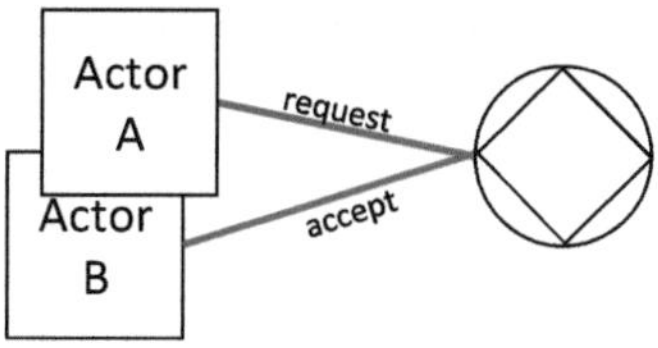

Fig. 5.6 ECF 6: Split request and accept initiator

5.1.1.7 ECF 7: Split Initiator ('Shattered Requests')

An initiating actor role is elementary in the sense that there is no smaller unit of responsibility, authority and competence for the initiation of a certain transaction. Such an actor role can also be a committee. It could very well be possible for the committee to become internally divided to such an extent that no decision is taken and the transaction takes unnecessarily long or does not take place at all.

For example: the legislation and regulation stipulates that a product must meet new requirements. The management group internally disagrees on the cost items and how the costs involved for the adjustment of the product in order to meet the new requirements should be divided. As a result the assignment takes so long that too little time is left to correctly build and test the product. Figure 5.7 depicts the way in which ECF 7 is represented.

5.1.1.8 ECF 8: Overlapping Initiator ('Double Requestor')

It may occur that a transaction is started by different initiating actor roles, without them even being aware of such. As a result the transaction is performed twice (costing unnecessary energy or leading to an undesired result), the executing actor role is confused ('your colleague already asked this, didn't he?') or the different initiators disagree ('that is my job!').

For example: you make arrangements with your partner to go to a pop concert. The tickets are 125 € a piece. At the entrance you find out that the both of you have bought tickets for the total value of 500 €. Figure 5.8 depicts the way in which ECF 8 is represented.

5.1.1.9 ECF 9: Missing Accept Initiator ('No Acceptance')

The initiating actor role initiates the transaction by placing an request; the executor promises the delivery and makes the product. However, when the product is delivered, nobody is there to receive it; it will therefore never be clear that this is actually the product that was ordered. Or the executor may think along the lines of 'no news is good news', feel that his work will probably be alright and therefore closes the order.

For example: someone orders a Smartphone via the Internet. Sadly this gets lost in transit. The supplier assumes that the mobile phone has been received and sends an invoice. However, as the customer did not receive the phone, he does not pay.

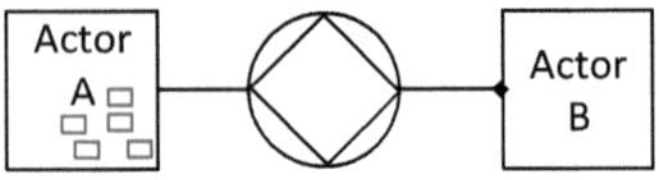

Fig. 5.7 ECF 7: Split initiator

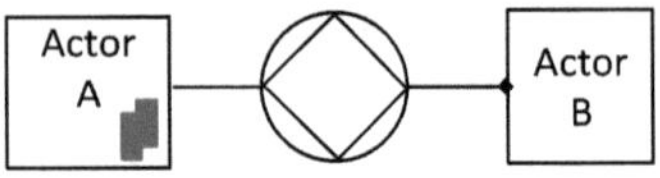

Fig. 5.8 ECF 8: Overlapping initiator

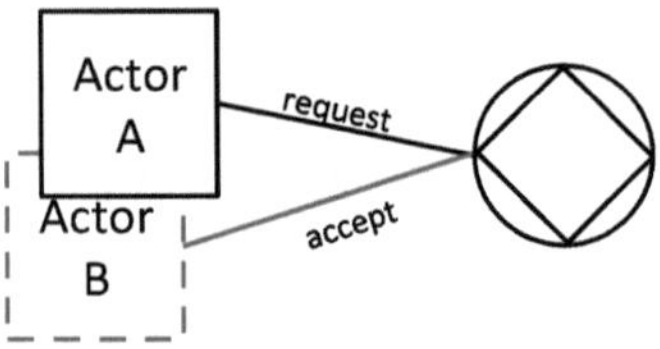

Fig. 5.9 ECF 9: Missing accept initiator

This results in a conflict. The supplier would have been better advised to change his business process and have recipients sign for the receipt of the (expensive) products. Figure 5.9 depicts the way in which ECF 9 is represented.

5.1.2 ECFs with Regard to the Role of Executor

5.1.2.1 ECF 10: Missing Executor ('Nobody Does the Work')

In practice, this Elementary Construction Flaw ECF10 is no different than ECF20—Missing transaction. The relation between a transaction and the executing actor role is after all per definition one-on-one. Without an executor there is no transaction. Still this flaw is listed separately because a missing transaction is not rarely caused by there being no actor role meeting the request. Figure 5.10 depicts the way in which ECF 10 is represented.

5.1.2.2 ECF 11: Unclear Executor ('From Pillar to Post')

Actor role A thinks that he knows who he needs to see to solve a problem. However, when he wishes to make a request, the executor cannot be found. The transaction therefore does not take effect.

For example: An employee discovers that his laptop adapter is broken. He calls the ICT helpdesk because he assumes that it is their task to solve this problem for him. The ICT helpdesk tells him that he needs to go to Facility Management. When he contacts Facility Management he is sent away: 'No, ICT does that, you need to contact the helpdesk.' The executor does exist in the construction (because colleagues with the same problem have been given a new adapter), but it is unclear who and where he is. Figure 5.11 depicts the way in which ECF 11 is represented.

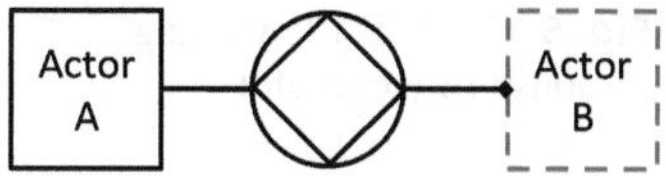

Fig. 5.10 ECF 10: Missing executor

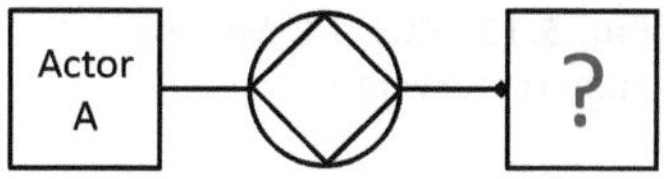

Fig. 5.11 ECF 11: Unclear executor

5.1.2.3 ECF 12: Missing Responsibility Executor ('No Responsibility for the Result')

As we said before an actor role can initiate a transaction without him having the responsibility to do so. The same applies to the actor role executing the transaction. The tasks have been divided without a specific intention of doing so (this has grown during the course of the years), but sadly nothing was formalized. As a result everyone just assumes that actor role B will do the job. Possible cause: the actor role is not linked to a position.

For example: a receptionist who has been working at the reception desk for years cleans the coffee machine every week. However a new receptionist does not feel responsible for the result (a regularly well cleaned coffee machine). She does not consider this to be a task for a receptionist. It is also not included in the job description; she has not been explicitly made responsible for the task. She only does half the work. Within 2 months the machine breaks down. Figure 5.12 depicts the way in which ECF 12 is represented.

5.1.2.4 ECF 13: Missing Authority Executor ('No Mandate to Do the Job')

Nobody likes to admit that he lacks a certain authority, certainly not within organizations. Actor roles are often happy to make promises ('of course I can help you'), but fail to act on them. Or sometimes the result of the transaction is only temporary as matters are reversed at a later stage. The actor role in question does not appear to have the authority to make the promise.

For example: Vitesse and Feijenoord are Dutch soccer clubs. A while ago Vitesse wanted to buy a Feijenoord-player. The technical director of Vitesse made an offer, which was accepted by Feijenoord. However, Feijenoord was told the next day that the deal was off: the technical director of Vitesse was in no position to offer that kind of money. Figure 5.13 depicts the way in which ECF 13 is represented.

5.1.2.5 ECF 14: Missing Competence Executor ('Don't Know How to Do It')

An actor role that has the authority and responsibility, but lacks competence, is a weak link. This can result in a bad product or no product at all.

For example: a project manager must arrange outsourcing to a company in the United Kingdom; something he has not done before. He loves the challenge and sets to work full of energy. After a while it becomes clear that the project is running

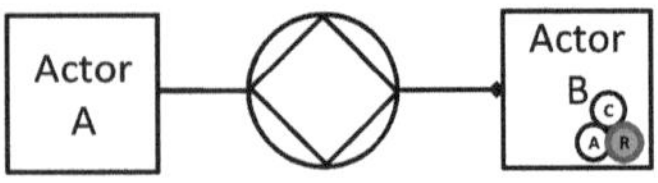

Fig. 5.12 ECF 12: Missing responsibility executor

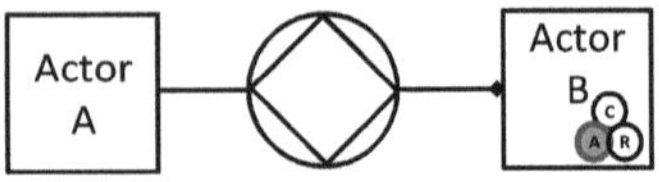

Fig. 5.13 ECF 13: Missing authority executor

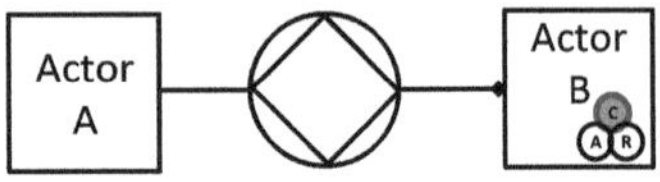

Fig. 5.14 ECF 14: Missing competence executor

behind on schedule. An internal evaluation makes clear that he has an insufficient command of the English language to be able to manage the external supplier correctly. Figure 5.14 depicts the way in which ECF 14 is represented.

5.1.2.6 ECF 15: Split Promise and State Executor ('Promise Is Not Kept')

We have seen that an initiator can split the coordination activities, i.e. request and acceptance, for example by delegating the acceptance to someone else. The executor can also split the activities that belong to his role (promise and state). This is not by definition a bad thing. It only goes wrong if the promise has not been discussed with the party making the state, for example because both parties apply different criteria.

For example: a bank has a sales employee who sells mortgages. This employee determines whether a client is eligible for a certain mortgage on the basis of a number of criteria. He promises a mortgage for a certain amount on the basis of these criteria. But the internal sales department—the department that actually processes the mortgage request and that later makes the state—applies different criteria. This could mean that the client is not eligible for the mortgage after all. It would have been more efficient if the sales employee had applied the same criteria as the sales department or was made responsible for the processing of the request, and fewer clients would be disappointed. Figure 5.15 depicts the way in which ECF 15 is represented.

5.1.2.7 ECF 16: Split Responsibility Executor ('Shattered Execution')

An actor role is elementary in the sense that there is no smaller unit of responsibility, authority and competence capable of executing a transaction. Sometimes the executing actor role is not elementary but split up, i.e. performed by a number of different people with different or sometimes even opposing interests. This can obstruct the transaction in question considerably.

For example: a project proposal must be approved by a steering group, but the steering group is unable to reach a decision. The reason: The meetings are always attended by so many parties that it becomes a divergent chaos of opinions. Everyone voices their opinion and do not want to be overruled; therefore the approval

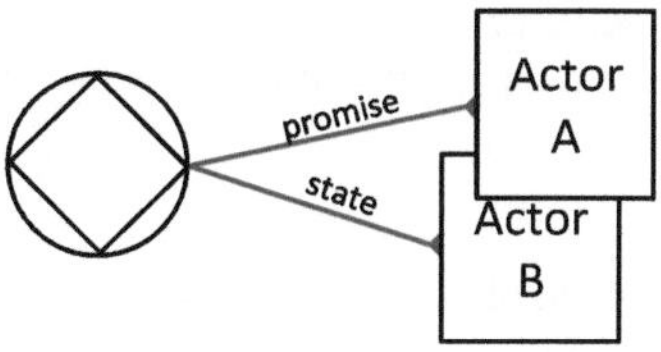

Fig. 5.15 ECF 15: Split promise and state executor

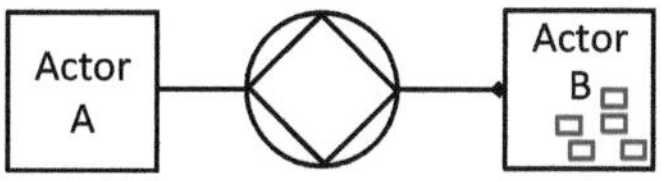

Fig. 5.16 ECF 16: Split responsibility executor

process takes too long. This organization would be advised to scrutinize the composition of the steering committee and to work with ambassadors/mandates rather than with so many different individuals. Figure 5.16 depicts the way in which ECF 16 is represented.

5.1.2.8 ECF 17: Overlapping Executor ('Double Work')

Sometimes there are multiple actor roles, all complying with the same request. This leads to work being performed twice, or different results; which is even more annoying. It would be good to consider which actor role would be best suited to execute the transaction. It shouldn't take long to determine which person is best suited to perform the task in view of the specific competences or the position that person holds in the organization.

For example: a team leader makes the request to draw up a list of the materials currently held by a department. Two people, actor role B1 and B2, both promise to do this, without being aware that the other made a promise as well. The lists drawn up by B1 and B2 show a difference in numbers. Actor role B1 draws up the only correct number, because only he has the right information. If, upon designing the cooperation, it had been clearly documented that actor role B1 was the only one to execute this transaction, no work would have been performed twice and no incorrect results would have been submitted. Figure 5.17 depicts the way in which ECF 19 is represented.

5.1.2.9 ECF 18: Insufficient/Missing Information Executor ('Missing Key Information to Do the Job')

The execution of a transaction (i.e. the production of something) is successful if the performing actor role has the right information. If information is missing or incomplete, this could lead to a bad or insufficient result.

For example: a steering group decides to approve a project proposal on the basis of an estimate of the costs and benefits (business case). However, the cost estimate is not based on past experiences and the benefits are terribly overrated. In the end the project turns out to have a negative return on investment. Figure 5.18 depicts the way in which ECF 18 is represented.

Fig. 5.17 ECF 17: Overlapping executor

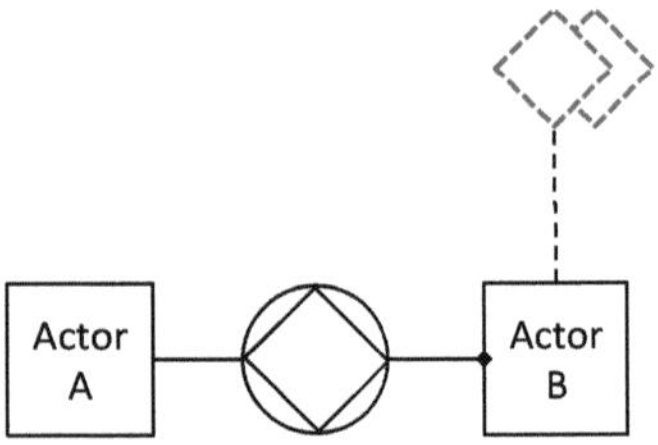

Fig. 5.18 ECF 18: Insufficient/missing information executor

5.1.2.10 ECF 19: Undesired Self-activation ('Undesired Own Initiative')

In a self-activating transaction the employee is the one who decides that an action needs to be performed, without having received an explicit request thereto. This means he is both the initiator and the executor. This can be intentional, for example if a logistics employee places an order when the stock goes below a certain explicitly required minimum level. However, it is often not the intention for the initiator and the executor role to be performed by one and the same person. In that case an explicit activation by another initiating actor role would be advisable.

For example: a contractor takes a decision on the construction beyond the agreed architectural framework, without first having his decision assessed. It turns out that his solution is not acceptable and the work must be redone. Figure 5.19 depicts the way in which ECF 19 is represented.

5.1.3 ECFs with Regard to Transactions

5.1.3.1 ECF 20: Missing Transaction ('No Awareness of Such Activity at All')

We sometimes assume that something has been done, but that is not always the case. This will cause problems later.

For example: the management secretary plans meetings with customers. However, the customers always arrive late. After a while it becomes clear that the secretary assumes that the customers travel by public transport and therefore failed to reserve parking spaces for them. Reserving the parking spaces is a missing transaction in this example. Figure 5.20 depicts the way in which ECF 20 is represented.

5.1.3.2 ECF 21: Unclear Transaction ('Unclear Agreements')

It is clear what the responsibilities of a person—the initiating actor role A—are, and the same applies for another person—the executing actor role B. A and B have agreed that B will deliver something to A. However the end result is not good, B turns out to have delivered too little. A and B reach another agreement and things go

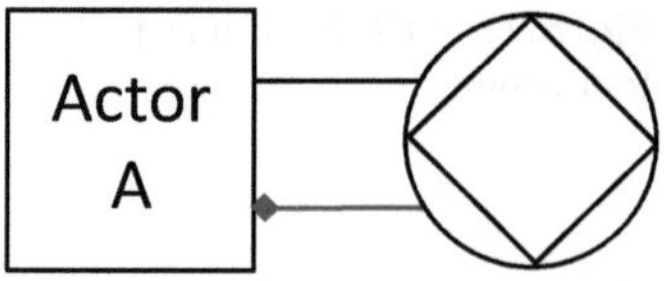

Fig. 5.19 ECF 19: Undesired self-activation

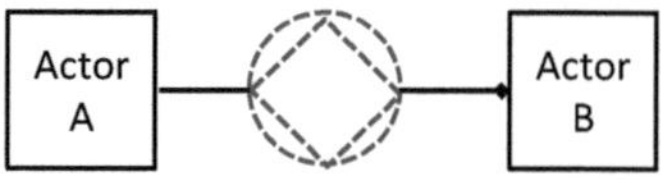

Fig. 5.20 ECF 20: Missing transaction

wrong again. What went wrong? A and B felt they made an agreement, but a closer analysis makes it clear that they have not understood each other correctly. This kind of miscommunication is not uncommon: we tend to talk at cross purposes when dealing with day-to-day matters.

For example: the job description of a, author states that he is responsible for supplying a manuscript. After a while it becomes clear that the writer does not concern himself with the images, just the text. His client (who drew up the job description) is of a totally different opinion. Figure 5.21 depicts the way in which ECF 21 is represented.

5.1.4 ECFs with Regard to Missing Information

5.1.4.1 ECF 22: Unclear Acceptance Criteria ('I Didn't Get What I Asked for')

A transaction always involves implicit acceptance criteria. In many cases this is fine, you don't need to specify everything. However, in certain cases it can be wise to list those criteria, certainly in cases where one could expect something to go wrong.

For example: when someone orders a pizza from a pizzeria, you don't need to explicitly add that the pizza must be hot and fresh. The employee of the pizzeria knows this.

But if the person requesting the pizza does this on his way home, he would be well advised to mention that the pizza must be delivered no earlier than 7 p.m., as that is the time he will be arriving home. Otherwise the pizza delivery man will find nobody home, will return the pizza to the pizzeria, while the initiator will still need to pay. Figure 5.22 depicts the way in which ECF 22 is represented.

5.2 The Approach of the Enterprise Engineer

Like any complete method, Enterprise Engineering consists of various components that complement each other. These are also referred to as 'ways'. Generally there are six ways [1], as shown in Fig. 5.23.

The most important ways, in terms of this book, are:

Fig. 5.21 ECF 21: Unclear transaction

Fig. 5.22 ECF 22: Unclear acceptance criteria

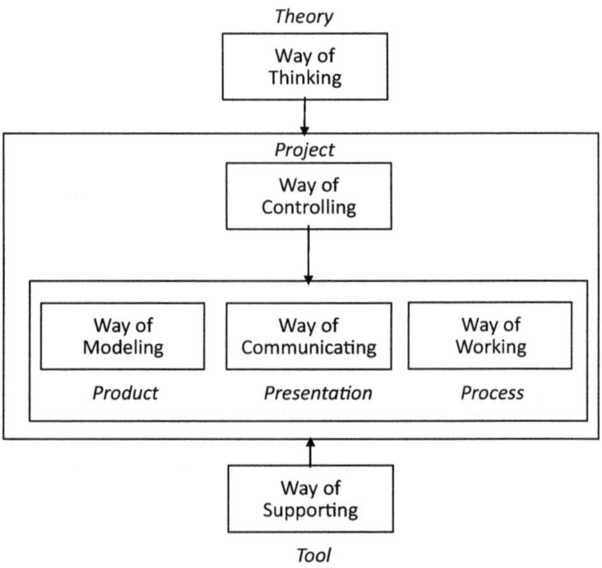

Fig. 5.23 The six ways of Wijers

- The Way of Thinking (WoT): This includes the theory on which the method is based, i.e. Enterprise Engineering.
- The Way of Working (WoW): The WoW describes the way in which work is performed in order to achieve certain results.
- The Way of Modeling (WoM): This is the DEMO©-modeling method, the completeness and coherence of the models.

We have discussed the WoT and the WoM in Chap. 4. We have not yet discussed the Way of Working. The WoW is the way in which the construction of an organization is exposed, even before this takes shape in Construction Models. After years of practical experience I have established the following WoW, which I regard as the most effective, logic and consistent.

5.2.1 Step 1: Determining the Scope

Before the current situation can be drawn up—the AS IS situation—we first need to know which area we must be consider. This area can be a single department, a project group, various departments and the contracts concluded between them, or the contact concluded by a department and the external environment such as suppliers et cetera.

But first we will need to make a different choice, one which will be included in the organization dimension. We need to choose between: governance, implementation, execution or improvement (see Chap. 4). Sometimes only a single dimension will need to be viewed, but it is also possible to view multiple dimensions. The choice fully depends on the problems faced by the organization and where they occur. The dimension chosen also strongly depends on the experience and insight of the engineer. However, this aspect of subjectivity is not such a problem. If an incorrect scope is chosen, for example a scope in which no or hardly any error patterns can be found, then it will become apparent soon enough once the interviews take place, or later, when the engineer is modeling and there appear to be hardly any Elementary Construction Flaws (ECFs). He then knows that he needs to adjust the scope. However, this seldom happens as the problems described by the customer organization already provide an indication as to which dimensions need to be looked into, even if the explicitly mentioned problem is only a supposition in this stage. Moreover, the experienced engineer will also know which problems are interconnected, and he will know what the connections are and what questions to ask, based on his experience.

In order to determine the scope, the engineer first has initial meetings and he studies the relevant documents such as process descriptions, job—and role descriptions, manuals and other relevant documents. Eventually the scope will include the concrete research area and the chosen dimension(s).

5.2.2 Step 2: First Inventory of the Research Area

5.2.2.1 Appointing and Interviewing Persons in Key Positions

Requested and provided documents can shed more light on the course of events in an organization or a part thereof. But it regularly happens that what it stated in documents, for example in job descriptions does not tally with the actual course of events.

Process descriptions, too, describe the ideal situation: they describe exactly how the process should take place in theory. The documents can be a way to determine the actual course of events. They are used as input for the interviews held with persons holding a key role. These persons should (try to) provide the answer with regard to the following issues:

- What does the process/do the processes look like?
- Which roles can be distinguished? What are the responsibilities, authorities and competences included in those roles?
- What are the production facts of the service provided?
- Who is responsible for which coordination activities? Who takes care of certain requests? Who makes the promises?

5.2.2.2 Determining the Story and the Ontological Facts

After a number of interviews a 'story' begins to take shape of the course of events at the department in question, between departments et cetera, and there is a (partial) insight in the business processes. As I said before: business processes are not elementary. The underlying transactions must therefore still be discovered. In the 'story' the engineer indicates the ontological facts, i.e. the agreements reached and the decisions taken. He needs those ontological facts to distinguish the transactions and actor roles.

5.2.3 Step 3: Drawing up the Spreadsheet AS IS

If the story is complete, the engineer draws up a clear spreadsheet which includes the following essential information:

- The area studied and all the transactions within this area
- The institutions, organizations, departments, persons et cetera. who/that start a transaction by making a request (the initiators)
- The institutions, organizations, departments that and persons who promise to comply with this request (the executors)
- The action taken by way of coordination activities and production activities. This involves requests, promises, states and accepts, but also: giving permission, giving the assignment, monitoring, drawing up (of reports et cetera.), deciding, taking care of, approving, assessing and outsourcing
- The level of the transaction: whether the transaction is datalogical, infological or ontological

The spreadsheet will contain a significant number of gaps. Some roles and functions may be missing. In order to complete this information, the spreadsheet will serve as a guide for the next interviews. If the information is collected in the spreadsheet, ECFs will come to light soon enough (for example, that there are no less than five initiators for one and the same transaction—a not unlikely overlap, as we have seen in practice).

5.2.4 Step 4: Modeling the AS IS-Situation

The AS IS situation is revealed with the aid of the Construction Model. This model contains the following elements:

- Composite actor roles
- Elementary actor roles
- Transactions
- Organizational boundaries
- Information banks

The AS IS model clearly marks all the ECFs as follows:

- Red question marks—the initiator, transaction or executor is unclear
- Two red squares within an actor role—there are different parties performing this actor role
- Lots of red squares within an actor role—responsibilities and authorities for the actor role in question are split
- Red dotted lines—an actor role, transaction or information bank is missing
- A red line with red dot near a self-activating actor role—this actor role is self-activating, which should not be the case
- Two red lines within an actor role—there are unclear acceptance criteria
- Two black actor roles as initiator whereby a red line has been drawn from each square in the direction of the transaction—there is an unjustified delegation of the acceptance to another actor role
- Two black actor roles as executor and two red lines which have been drawn from the transaction to the two black executors—the role of the executor is incorrectly performed by different actor roles
- A red circle with a 'r' in an actor role—this actor role misses the responsibility to initiate or perform a transaction
- A red circle with a 'a' in an actor role—this actor role misses the authority to initiate or perform a transaction
- A red circle with a 'c' in an actor role—this actor role misses the competences to initiate or perform a transaction

5.2.5 Step 5: Modeling of the TO BE Situation

In the *TO BE* situation all the ECFs have been resolved. It can be that only minor changes are implemented, such as the addition of information banks or the introduction of a couple of new actor roles. It is also possible that whole departments need to be reorganized; that for example many actor roles are added and that existing actor roles are moved to other departments. The new situation fully depends on the number of ECFs and the problem and the goals of the customer organization.

In the cases I use fragments of the Construction Models to illustrate the cooperation and certain ECFs. The composite actor roles (CA), elementary actor roles (A), transactions (T) and information banks (APB) seem randomly numbered, but they are not. If, for example, only A01 and A34, or T02 and T29 are present in a Construction Model, this means that the other transactions and actor roles have been indexed, but are not present in this fragment of the Construction Model.

5.3 Applying Enterprise Engineering at Random?

Just image, if you will, a car of which the construction is unknown. The construction cannot be retraced, as there is no hood that can be opened. It smells of gas once in a while, there is smoke coming from the dashboard, the brake sometimes does not work, and every now and then the vehicle produces very odd sounds. You provide a diagnosis and put this to the mechanic. The mechanic explains that you should repaint the car and that you should take it to the car wash at least twice a week. This does not resolve the problem. You go back to the mechanic. He changes the tires and tells you that this should solve the problem. But alas the problem is not resolved.

The same applies to organizations; the old paradigm claims that the exterior of the organization is visible, while the interior is not. Enterprise Engineering changes all this. There is a hood that provides access to the interior of the system. The method is the tool which allows the hood to be opened. And just take a look inside: there are all parts that work together. There are actor roles and transactions. What is wrong? It abounds with Elementary Construction Flaws (ECFs). Once those have been repaired the organization will again run smoothly—like a car that finally runs without problems. Are we done? No. Because there is still the question where you are headed? Let's take a look at the destination first. Then we can determine whether the construction of the car needs further adjusting, allowing you to arrive there without problems.

That is what the next chapter is all about: where can you apply Enterprise Engineering?

Reference

1. Seligmann, P. S., Wijers, G. M., & Sol, H. G. (1989). Analysing the structure of IS methodologies: An alternative approach. In R. Maes (Ed.), *Proceedings of the first Dutch conference on information systems*. Amersfoort.

The Scope of Enterprise Engineering 6

Although Enterprise Engineering provides a lot of insight into the organizational construction, the insight is never a goal in itself. Insight in the organizational construction is the way to track and correct, for example, Elementary Construction Flaws (ECFs—see Chap. 5). Optimization of current business processes is however just one of the applications. Having an insight into the organizational construction can also help achieve better results in various other organizational areas. In this chapter I will discuss the most important ones: mergers and takeovers, reorganization, complexity control and reduction, (the implementation of) ERP packages and the implementation of methods. The common denominator of these areas of application is that it involves organizations on the move; organizations that are partially or fully subject to change. Enterprise Engineering can then be used to introduce these changes and adjustments correctly in one go, without wasting time and money on experiments—and the trial and error that comes with them—the effects of which are not clearly defined beforehand.

6.1 Mergers and Takeovers

Mergers and takeovers turn two organizations into one single organization. Enterprise Engineering can reveal the constructions of both organizations for the purpose of sliding them together. This provides an objective approach and a way to realize the best possible connection. But still this method is seldom used in practice.

There is generally little difference between a merger and a takeover. What's called a merger usually appears to be a takeover and each takeover involves a merger. Takeovers usually involve some more inequality in terms of the organizations' size. The strategy, the policy and the goals of the organization taking over are fixed and are leading. The party that is taken over often takes second place. Obviously, that does not mean that the policy and the strategy of the party being taken over have suddenly disappeared. One way or another the policy and the

T. Janssen, *Enterprise Engineering*, Management for Professionals,
DOI 10.1007/978-3-319-24172-2_6

strategy of both parties must be combined in order to create one single organization with a uniform strategy and policy. There are different ways to achieve this.

The easiest option is a kind of absorption. The party taken over (B) is fully absorbed by the party taking over (A). The staff of B will then need to re-apply for a job and may be assigned new positions. This is usually the scenario if A is mainly interested in increasing its market share and in doing so wishes to attract the customers and possibly the top employees of organization B.

Some organizations will opt to leave organization B fully intact. This means there is no or hardly any integration. This was the case, for example, at ABN AMRO's takeover of the Italian bank Antonveneta (an Italian bank). This initially leads to no or hardly any problems because B can continue to work as before. It only becomes a problem if the company taken over takes strategic decisions that are not in line with the strategy adopted by the parent company.

In other cases, A might, for example, be interested in expansion and introducing a product or service in the market for which it requires all or part of B's expertise. Also then is it possible for B to operate autonomously. But usually a process will be set in motion to achieve a certain level of integration. If there is no insight into the black box, the constructions sadly remain hidden, and often trial and error tactics are applied because nobody really knows. Functions that appear similar and parts that seem the same are incorporated in new departments. Things are almost always 'resolved' in a way that is based on the hierarchical structure, i.e. the organizational chart. Such always leads to problems.

Let's suppose that a project management department is set up or expanded. After all, project managers do the same work everywhere, at least that is what people tend to think. However it can become clear pretty quickly that the position of project managers in organization B involves totally different tasks than that of organization A; a project manager at A can perform a total different set of elementary actor roles than the project manager at B.

It is for example possible that the position of project manager at A is exclusively aimed at project management and that the project managers at B are working at a more specialist level because they have these skills. If these different project managers need to work together in a new or existing department, this will naturally result in conflicts. It is therefore no surprise that many employees leave the company if the company is subject to a merger or takeover. Usually the best employees leave first.

The problems arising in the event of a takeover are often thought to be 'company culture'-related. This may be true, but we have seen that the existence of Elementary Construction Flaws (ECFs) causes certain behavior, and it is also possible that this behavior stands in the way of the integration of the two parties. It is therefore possible, once again, that the heart of the problem lies in that construction. By the way, it is not my intention to trivialize the effects of major cultural differences; those can indeed cause mergers and takeover to fail. I just wish to emphasize that problems within organizations are not always caused by cultural differences; they can also be caused by compensating behavior caused by ECFs. The project manager at A is asked to initiate a chain of transactions that are not based on specialized decisions, while those at B are. The manager at B will accuse A for doing too little in specialist terms, while A will accuse B that he is involved in matters he should not be involved in, because that is not how we (the party taking over) do things! Naturally these issues could just as easily arise in other positions, requiring the party taking over to do its utmost to correct it, if at all possible.

Having insight in the construction offers an opportunity to ask questions and receive answers, such as: what do our project managers do (in terms of elementary actor roles), what are they supposed to do (which transactions do they initiate and which do they execute), and do they have the right authority, competences and responsibilities? Do we need to employ more project managers or do we have too many in light of our strategy? The latter will not be determined on the basis of the capacity of the organization, but on the basis of the construction. Having insight into the construction is therefore a useful tool during mergers and takeovers, and it enables one to make decisions based on facts rather than assumptions. The ECFs present will become apparent in the Construction Model. It is by the way most probable for cultural problems to arise during a merger or takeover. Still the construction is the basis for analysis. The construction happens to form the best basis for dealing with any cultural problems as well. We will discuss this in more detail in Sect. 8.1.

6.2 Reorganizations

Enterprise Engineering is also of significant value for reorganizations. If one has insight in the construction of the organization, one can adjust the organization effectively.

Society is always on the move, as are organizations. Reorganizations occur all the time. Almost every organization feels that their quality should improve and that they can and should be more effective, flexible and efficient. And that is true. However, in times of economic crisis these seldom involve reorganizations that will really lead to intrinsic improvements. In such a situation, organizations are just looking for ways to do the same work with fewer employees. The word 'reorganization' has in those instances just become a euphemism for 'wave of redundancies', and reorganizing is just the way to become cheaper, as employees are expensive. For that very reason, often the wrong decisions are taken, negatively affecting the quality of the cooperation and therefore the quality of the product or service. The reorganization then achieves the opposite of what was intended. The results of those wrong decisions will only come to light later, and sometimes these are not only negative, but even disastrous. Only a thorough understanding of the complex system of the organization, hence the use of Enterprise Engineering, will help to solve the problem.

Any organization that deals with its Elementary Construction Flaws (ECFs)—a thorough MOT inspection as it were—will become much more efficient. Still sometimes more is needed. For example, a change of strategy is a good reason to reorganize. A logical step in that case would be to compare the current organizational construction with the strategy. The key question in those cases will be: can my organization handle the adopted strategy or do the businesses processes need further optimization first? After all, each construction leads to options, but also to limitations. Sadly, organizations lack patience during times of reorganization. Everything has to become faster and more cost-effective. They only focus on the strategy and positions are somewhat impulsively merged or left out altogether, without even having the slightest objective idea of the AS IS situation. Hired staff is quickly regarded as redundant, while it may very well be that they have the expertise that allows the cooperation to run smoothly.

The decisions taken during a reorganization that have major impact on the construction of the organization, are compromises of all sorts of different interests. It is obvious that the plan that finally takes shape is a combination of all those different interests. It remains to be seen to what extent this plan will actually benefit the organization as a whole. Time will tell what the black box will produce. The organization will lose people it would have liked to keep, while those the organization would have rather have done without will stay. Managing a company on the basis of interests rather than facts is a recipe for disaster. Just fiddling about in a complex system without having a real clue is never without consequences.

However Enterprise Engineering may help to resolve the issue of managing on interests and the games of politics; it is possible to expose this so-called 'informal organization' (see Sect. 8.4).

Only when one is aware of the ontological level of the organization and compares it with the strategy will it be possible to look into which positions, departments and project teams are no longer required and determine which new parts would be useful. In short, only then the organization can be rebuilt. This is something completely different than removing teams or staff departments at random, for the simple fact that cuts need to be made. Often the teams responsible for *quality* or *improvement* will suffer: the teams that should achieve the various departments working in the same way. As soon as they appear to have reached their goals, the teams are dissolved. Sadly, later on these departments often fall back to working as they always did, something that was intended to prevent in the first place. Then the teams responsible for quality and improvement are called in and can start from scratch again. It is this cycle of introduction, dissolution and reintroduction that wears employees out and who can blame them? The remark 'management has dreamt up something again' is a much heard and justified complaint. Employees understand that the actions that are being taken, however much they are promoted, are only temporary.

6.3 Outsourcing

It is not rare that a reorganization involves outsourcing. Outsourcing is still considered the miracle drug that will keep costs to a minimum. However, research has shown that it is extremely difficult to actually be cheaper with outsourcing as it generates lot of overhead costs [1]. Only organizations in certain sectors manage to increase their return on investment with outsourcing, for example in certain areas of retail. In spite of that, in certain cases it is still wise to outsource activities. Continuity can be a valid reason. If, for example, a certain ICT system must keep on running and no experts can be called it, this expertise must be found outside of the organization, sometimes even in other countries.

Companies apply outsourcing without having insight in the construction. Cuts are made within the own organization and sometime entire departments disappear. This results in a gap and therefore a new organizational boundary. The external party will need to bridge this gap seamlessly if the business is to continue running smoothly. This requires effort, which is why the earlier mentioned Service Level Agreements (SLAs) are drawn up.

An SLA details the services to be provided, as well as the rights and duties of the service provider and the client (in this case the newly appointed external supplier and the own organization respectively) with regard to the agreed quality level of the services and/or products to be provided. The purpose of an SLA is to clearly document the agreements and forms the basis of the cooperation between the provider and the client. All the agreements and expectations are explicitly documented in the hope of optimizing the cooperation. This is a valid goal. But why is it that this cooperation so often fails? Often the parties involved in the outsourcing end up in court.

It becomes clear in practice that the outsourcing organization is seldom able to provide a clear and unambiguous description of the expectations. If a client is unable to clearly define what they want, how can a supplier meet these expectations? This is not, by definition, someone's fault, but it is due to the language used in SLAs. The description of the 'Service' in the SLA is open for interpretations in different ways; the volume of the service provided may be unclear or the availability and reliability thereof are unclear or insufficiently documented. In short, it is unclear on the basis of which criteria the client accepts the product or service. If the cooperation is disrupted and certain activities are not performed, people will tend to show compensating behavior: the familiar 'if you don't do it, I will'. Or accusations are made ('they want something different every time'), negatively affecting the cooperation. And many other issues may arise, as I have explained in the description of the ECFs.

By applying a solution that is geared towards the construction you avoid having errors and ambiguities within the cooperation. The SLAs are then described in terms of actor roles and transactions. After all, the service provider and the client are two composite actor roles which need to be elaborated further and split into elementary actor roles. The cooperation is not only one transaction between the two composite actor roles (although it may appear so at first glance on the basis of the SLAs) but a complex collection of countless transactions between elementary actor roles. That collection must therefore be uncovered if one wants to optimize the cooperation. Which requests are made precisely and by whom to whom? Who promises something and who accepts what? SLAs are too ambiguous for a description of the construction of the cooperation in practice.

6.4 Complexity Reduction and Control

The greater the role of ICT in businesses, the more the complexity plays a role in the complete application landscape. A part of this complexity cannot be avoided; large organizations with countless departments and employees form a complex system, whether we like it or not. However, a large part of the complexity in applications can be avoided and certainly made controllable. In this section I will discuss two types of complexity: *technical* and *functional* complexity. We will see that these types, contrary to popular belief, are mainly caused in the *governance* dimension (see Sect. 4.8 for an explanation of the four dimensions).

Let's first define what 'complexity' is exactly and what is meant by the reduction and control thereof.

Complexity can best be determined with the aid of the system theory. As we already discussed in Sect. 3.1, a system consists of four components: composition, structure, production and environment. The elements in a system are also referred to as 'nodes'.

Complexity in the system can be due to:

- the number of nodes;
- the number of connections;
- and the dynamics: how often and how fast does the composition change.

The road network is a specific example of a system. The nodes are for example the cities or other places where the roads interconnect. The roads are the connections. The dynamics include the changes (and their frequency and speed) in the network of roads and traffic nodes, such as the breaking up of old roads, detours and the construction of new roads. If there are road works being performed in 90 places in the Netherlands, the road system is more complex than if road works are being performed at only 40 locations. Taking a detour usually involves travelling more roads, thereby increasing the chances of mistakes, such as taking the wrong exit et cetera.

Let's assume that an organization uses multiple ICT applications and the decision is taken to remove the oldest application. The first thing that will need to be done is to determine the connection with other applications. Sometimes there are tens of thousands connections with other applications! After months or years, the 'plug' can finally be removed from the old application. All the connections have been rerouted to the application(s) that are continued. This seems to be a type of complexity reduction. After all, one element, i.e. node, has been removed. But in this example more than 10,000 connections must be rerouted to other applications to allow for the removal of the old application. There might be one less node, but the number of connections has grown substantially. There is a major risk that the system has become more complex. I am referring to technical complexity here: the complexity of the system itself. Only dealing with this type of complexity does not guarantee success.

Technical complexity becomes visible during the maintenance of the system: it can turn out to be very difficult to add an application, or changing an application can take a lot of time.

Another form of complexity is experienced by the users of ICT systems: the functional complexity. A system is functionally complex if the employees need to for example open three different applications in order to enter customer details. That functional complexity is even more substantial if such customer information is presented in different ways in those applications.

A familiar application of complexity reduction is *application rationalization*. In the 1980s and 1990s the focus lied on decentralization; as a result all sorts of departments and business units started realizing their own ICT solutions in order to support business operations.

New techniques were purchased and implemented all the time in order to be as innovative as possible and be ahead of the competition. This resulted in a sort of 'island automation', i.e. an application landscape showing various kinds of local technologies. This only become a problem when a wave of mergers and takeover followed in the 1990s, causing the new organizations to be left with a complex application landscape that consisted of (shaky) additions of countless technologies. Currently, there is hardly any time to clean the redundant systems. The result: approximately 80–90 % of the ICT budget is used for the maintenance of these systems [2]. These costs also mount up each year due to the increasing number of failures. At the same time the flexibility of the landscape is reduced, which means that it is very difficult to add new systems and phase out the old ones. As a result, only a small part of the innovation for which the ICT support had been set up can actually be realized.

It should be clear by now that an organization benefits from a *rationalization of the application landscape*. By this we mean: using applications for a specific purpose and making the provision of information more efficient in order to support business operations. Reducing the complexity forms an unmistakable part of this. As we have now learned that the construction of an organization consists of elementary actor roles and transactions, we have an anchor point. If ICT should perfectly align with the cooperation, then it should establish an exact alignment with to the actor roles and transactions present. Enterprise Engineering allows you to exactly define the roles of the application and those of the actors in the business. The Enterprise Engineer draws up an AS IS Construction Model of the organizational part where the applications run. He then visualizes which applications support which transactions. It will quickly become clear which applications are unnecessary, for example because they do not support any transaction. The end result is a *migration strategy* on the basis of Enterprise Engineering. A migration involves adding certain functionalities to other applications and withdrawing current connections, allowing the removal of an application.

In short: in order to avoid technical and functional complexity, one needs a good migration strategy based on a factual and objective analysis of the construction of the target organization where the applications run and that of the application landscape itself. That migration strategy must be based on the efficient use of all the applications in the business operations.

The major benefit of good migration is an enormous reduction of the total cost of ownership: the costs of running and maintaining the systems. A large government organization in the Netherlands was able to phase out 49 % of all the applications in this way, lowering the cost by more than 49 %. This often involves a cost-saving of several millions. Developers also benefit from a successful migration. It is much easier and cheaper to integrate a new application in an existing system that has been rationalized.

Organizations that only look into removing technical complexity during a complexity reduction are effectively wasting their time and money. The cause of technical and functional complexity lies largely in the *governance* dimension. Examples are decisions taken on the basis of insufficient information, a lack of insight into the consequences of a decision, lack of focus on complexity and an unbalanced division of responsibilities, authorities and competences.

Only when the *governance* complexity is under control, can the technical and functional complexity be effectively addressed. One needs to tackle the problem at the source before dealing with the consequences. How this is achieved by using Enterprise Engineering is described in the Complexity control case (see Sect. 7.5).

6.5 ERP Packages

Nowadays almost all organizations use ERP packages to support their business processes. Enterprise Resource Planning has largely replaced separate software applications and should offer a total solution to the users. Generally speaking an ERP system consists of separate modules. Each module supports its own process or task, but is integrated into the bigger picture. A benefit of these packages is, for example, that they use one single database, which means that data, such as client data, only needs to be entered once. ERP systems should eventually lead to better user-friendliness, efficiency and cost-effectiveness. But sadly great problems arise soon after the implementation, causing the costs to mount up. This is avoidable if the construction of an organization is placed next to and compared to the construction of the ERP package.

Quite a lot of work needs to be done before the system connects well to the organization. In order to realize that connection, three options are available for the implementation of the selected ERP package:

- One accepts the ERP software as leading and adapts the organization to align with the ERP software
- One accepts the organization as leading and adapts the software to align with the organization
- One chooses the middle road, whereby the ERP package and the organization are both partially adapted

Most organizations opt for the middle road, whereby they, for example, adjust their own processes and build their own modules to the system in order to create a good connection. Still, the system will continue to be inadequate, it keeps on failing to meet the requirements of the systems users. There are only a few ERP packages that are not cursed in increasingly strong language.

Because organizations where this problem occurs still don't look in the black box, they have no other alternative but to complete a long implementation project of trial and error. They make changes in the hope that the package will eventually turn out alright. That this almost never happens is really not surprising. If one does not *exactly* look at what the user does (so at the transactions this user is involved in), and at the support the application provides for the user, this faulty method of implementation will never lead to a match.

If the user (A) is a financial administrator, he has a composite actor role based on that position. The organization knows the financial administrator has various roles, but is unaware of the elementary roles contained in that position. Let alone that it is aware of all the transactions initiated or executed by the financial administrator. The software developer knows that a module must be made for a financial administrator and develops an application (Z) for certain tasks required for that position. Of course the software company is not aware of what position A precisely entails at organization X. This means that once application Z is delivered, this product will need to be adjusted, leading to additional costs. But still: if both parties only partially understand what the position of financial administrator involves, and therefore what module Z is required to do in the ERP package, that module—despite the adjustments—will never fully meet the requirements of A, who is the one using this module. And therefore those working with the software remain dissatisfied.

The solution is fairly simple, if you think like an Enterprise Engineer: the roles in the package should be aligned with the roles of the users. A composite actor role as 'financial administrator' must be examined closer in order to determine which elementary actor roles it contains and which transactions are needed. The same applies for the package or specific module. This too has a construction. The elementary actor roles of the module 'financial administrator' must be modeled. After which it must be determined where the connection is not optimal. The organization then needs to decide whether the roles in the software package needs changing or that the roles performed by the financial administrator need to be changed.

An example. Let's assume that the inspection of the annual budget is a self-activating actor role in the ERP system, this means that the inspection is started by the ERP system without an explicit request thereto from a human actor role. But what if that same inspection in the organization is only allowed to be started after a colleague in a certain actor role has explicitly made a request thereto to another actor role? In that case the first actor role initiates the transaction 'annual budget inspection', but this is not in line with the system, as in the system the initiation is automated. This is just one example of the countless mismatches that can occur. Many of these problems are not or seldom discovered. People do complain about the ERP system and about the problem that things are not possible or are not as the organization would like them to be, but the underlying causes (in terms of ECFs) remain hidden.

6.6 Implementation of Methods

Many methods worldwide were specifically designed to deal with a certain problem. Often the popularity of a method is not due to its proven effectiveness, but rather to a hype surrounding it. Lean, Agile, ITIL, Prince2, these are all practices, work methods that are or are not successful. In the ideal scenario these methods are a best practice: a proven effective approach.

Whichever way you look at it, each method in itself is a simplified version of the perceived reality and has little to do with the objective reality. The methods are therefore rarely suitable for direct use. They are models. A model of an atom, for example, with an atom nucleus that consists of protons, neutrons and electrons, is also a simplified reflection of reality; an atom does not really look like that. It is a way to simplify reality in order to show what it looks like. But the atom model is based on a thorough scientific substantiation.

This does not apply to the most practices and best practices, not even if the latter becomes the norm (the standard: a widely acknowledged and applied work method).

Even if the model or method is the result of years of scientific research and is closely linked to the objective reality, even then one should be careful when implementing it. The reasons for this are: (1) the effect on the organization is not yet known, (2) methods are adjusted to suit specific local circumstances in practice, (3) the order practice-best practice-standard is reversed, and (4) a real implementation has not yet taken place.

6.6.1 The Effect on the Organization Is Not Yet Known

As I said before: without Enterprise Engineering there is no insight into the construction of the organization. This causes problems when changes are implemented. When a new method is introduced, for example, this causes the arrival of new elementary actor roles and the disappearance of old ones. When introducing Agile/Scrum (a flexible method to make software products) the role of Product Owner is introduced. This Product Owner has a certain set of tasks (by the way: a 'Role' in Agile/Scrum is not the same as an actor role in Enterprise Engineering). In principle the introduction of new roles is not a problem. Sometimes Elementary Construction Flaws (ECFs) in an organization are caused by missing actor roles. These are gaps in the organizational construction which are filled up by the introduction of Scrum. However, if there are no visible gaps, or only partial gaps, then the Elementary Construction Flaw of overlapping actor roles can arise. The organization without Enterprise Engineer has no idea of the organization's construction, or of the construction of Agile/Scrum. Measures for changes are taken without knowing the effect it has on the organization. The organization will then run the risk that all sorts of construction errors arise. The more overlapping actor roles there are, the more FTEs and the greater the chance that people stand in each other's way. This means that the organization will need to Scrum very hard (which should result in a cost saving) if they wish to recoup this money. Eventually the implementation will escalate and the method is blamed; after which it is removed. But the problem is never or hardly ever the method itself. For example, it has been proven that the introduction of ITIL (IT Infrastructure Library), a set of best practices meant to perform the maintenance of ICT systems in a controlled manner, in fact resulted in the increase of the Total Cost of Ownership [3].

6.6.2 Methods Are Adjusted to Suit Specific Local Circumstances in Practice

The introduction of new methods makes employees inventive if they feel that their job isn't getting any better. They will then use the methods at their own initiative and on the basis of their own ideas. In practice, work is no longer being performed with, for example, the pure form of Agile/Scrum or Prince2, but with an approach that is the original method in name only. The adjustments are made with good intentions, but those fiddling with the method often have no idea what they are doing and what they are ruining, because this too takes place without knowledge of the underlying Construction Model. This means that all sorts of adjustments are being made with all sorts of unforeseen consequences.

6.6.3 The Order Practice-Best Practice Standard Is Reversed

Time is money and the management therefore wants results sooner rather than later, and this also applies to the introduction of a new method. When a method is introduced in phases, one starts with a practice. If this approach appears to be successful, one evaluates whether it can also be introduced to other projects, departments or companies. As times goes on the introduced method can be promoted to best practice and others who wish to introduce the method can do so. If the approach repeatedly proves its effectiveness, the method become the norm, the standard and becomes the way to work.

In large organizations this process can take up 2–3 years. As time is often short, many will opt for a short-cut: a direct implementation of the method as the *standard*.

This is a reckless course of action that never works. Every manual warns for this, as each organization is different—but it seems that chapter is never read. The management thinks: 'Others have given this much thought, so it must be working'.

The number of Elementary Construction Flaws (ECFs) resulting from such a forced implementation is countless.

6.6.4 A Real Implementation Has Not Yet Taken Place

What is implementation? Some people think that an implementation of a method requires nothing more than a few presentations. They have a too optimistic view of an implementation. Their idea is: 'After the presentation on the method has been given, everyone will surely start to use it independently'. It is a nice idea, but that's not how it works in practice. A method is only implemented once everyone thinks and works as prescribed by the method. If an organization wishes to implement method X, coaches need to be appointed or hired to help people on the work floor to work according to the principles of that method. Only then can the implementation be realized and the direct problems can be determined. Implementation is a field of expertise that required specific competences. Changes do not come about automatically.

Naturally a thorough advance research is required in order to determine whether a certain method can be implemented at all given the current situation and organization. Sometimes the steps that need to be taken are simply too large. If a method is considered in view of saving costs, it must of course be assessed whether those savings can actually be realized. Many change managers are so convinced that a method will work that they are blind to the possibility that it may not. Adopting a reasonable attitude without a too radical belief in the perfect method can save an organization a lot of trouble.

References

1. Gartner. (2003). *Controlling IT outsourcing costs.*
2. Braet, A. (Ed.). (2000). *Taalbeheersing als communicatiewetenschap. Een overzicht van theorievorming, onderzoek en toepassingen*. Bussum: Coutinho.
3. Derksen, B. (2004). De volgende stap voor ITIL. *Praktijdgids De Controller en Informatie-management, 42.*

A Number of Cases

7

In this chapter we will discuss a number of cases to illustrate how Enterprise Engineering works in practice. But first we will look at the cooperation between software architects and software developers within a department, and then the cooperation between a development department and an ICT Operations department, followed by a case on purchasing building materials. This is followed by a case on complexity control and then one concerning the implementation of Agile/Scrum in a large organization. The last case is one involving a reorganization of a government body.

The cases provide examples of the issues experienced by organizations and how these issues can be resolved with the aid of Enterprise Engineering. The descriptions of the examples are by no means complete; it would be impossible to include a complete project within this book. The cases have been anonymized on request of the client organizations.

7.1 An Example Case as Refresher

The cases in this chapter often use the Construction Model to reflect the cooperation. This paragraph is meant as a refresher of that what we have already discussed in the earlier chapters and which may have slipped your mind. It concerns the cooperation between the fictive companies Telco and TOP. This Construction Model does not contain Elementary Construction Flaws (ECFs—see Chap. 5). I will explain exactly how the Construction Model has been set up.

7.1.1 Brief Description of the Situation

Telco is a small company that sells mobile phone subscriptions. The company has outsourced the stock management of the mobile phones to TOP, an external party. TOP makes sure that the stock is kept at an adequate good level, that Telco's

T. Janssen, *Enterprise Engineering*, Management for Professionals,
DOI 10.1007/978-3-319-24172-2_7

customers receive mobile phones and that broken phones are repaired by the manufacturer.

7.1.2 Composite Actor Roles

The gray rectangles and squares in Fig. 7.1 are the composite actor roles. CA is the abbreviation of composite actor role. The 'stock manager', for example, is responsible for far more transactions than just the delivery of a mobile phone. As it is not important to show these separate transactions in this example (and therefore also not the elementary actor roles that go with it) we have chosen to reflect this actor as a composite actor role. This reflection (the gray square) does show that he, in his role as 'stock manager', is responsible for more transactions or that it is irrelevant for this case to know the underlying elementary actor roles.

CA12: Customer
: The 'client' purchases a service, in this case a mobile phone subscription including a new phone.

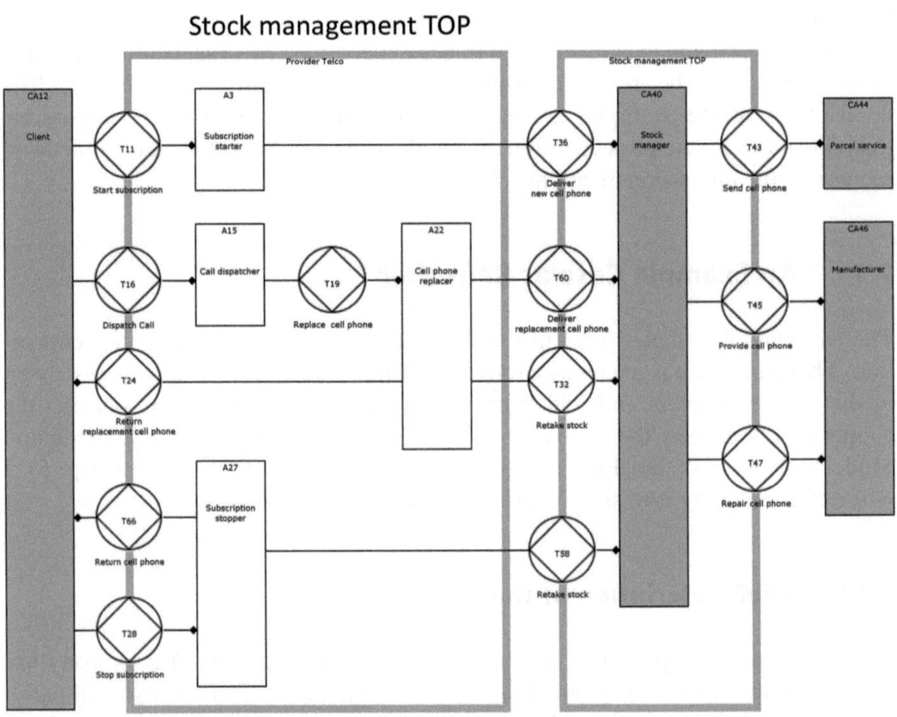

Fig. 7.1 Construction model of stock management TOP

CA40: Stock manager
The 'stock manager' is responsible for the transactions concerning mobile phones, such as keeping the stock at a certain level.
CA44: Parcel services
The 'parcel service' is for example a company like PostNL. They send the mobile phones to Telco's clients.
CA46: Manufacturer
The 'manufacturer' manufactures the phone.

The titles of all the actor roles and transactions are written between inverted commas.

7.1.3 Elementary Actor Roles

The elementary actor roles are indicated by a white square or rectangle. These are:

A3: Subscription starter
The 'subscription starter' is for example responsible for the registration of new clients.
A15: Call dispatcher
The 'call dispatcher' receives a phone call the moment a customer needs a new mobile phone to replace his old one.
A27: Subscription stopper
The 'subscription stopper' registers for example the termination of subscriptions.
A22: Cell phone replacer
The 'cell phone replacer' arranges the replacement of a mobile phone and contacts TOP about this.

In comparison: The composite actor role 'stock manager' is involved in different transactions, but for example the 'subscription starter' is only involved in the execution of one single transaction, i.e. the commencement of the subscription.

The title of an elementary actor role is usually not written in particularly beautiful English. What's important is that the title provides a description of the actor role that is as precise as possible. If the transaction is starting a subscription, then this transaction is called 'start subscription'. The elementary actor role that goes with that will logically be given the title 'subscription starter'.

Please note: the various elementary actor roles are not necessarily awarded to different persons.

The 'subscription starter' and the 'call dispatcher' are two elementary actor roles that require specific responsibility, authority and competence. This does not mean that they cannot be performed by one and the same person.

The tiny black dot (or rather small diamond shape—see again Fig. 7.1) indicates that the actor role (composite or elementary) takes care of the execution of the transaction. The coordination action *promise* is therefore provided by this actor role. If no black diamond is indicated with an actor role, this actor role is the initiator, who starts the transaction with the *request*.

7.1.4 Transactions

The circle represents the four coordination activities: *request*, *promise*, *offer and accept*. The diamond represents the production fact.

T11: Start subscription
T16: Dispatch call
T24: Return replacement cell phone
T66: Return cell phone
T28: Stop subscription
T19: Replace cell phone
T36: Deliver new cell phone
T60: Deliver replacement cell phone
T32/T58: Retake stock
T43: Send cell phone
T45: Provide cell phone
T47: Repair cell phone

7.1.5 Organization Boundary

The organization boundaries have been indicated with a gray box. These are usually departments, but can also refer to a complete organization.

Provider Telco: The telecom provider Telco.
Stock management TOP: The external party that is responsible for ordering cell phones and having them repaired.

7.1.6 The Construction Model

Provider Telco receives the request from a 'client' to start a new telephone subscription. The 'subscription starter' executes this request. As Telco outsources the delivery of mobile phones to the external party TOP, the 'stock manager' there is contacted. The 'stock manager' at TOP is in turn involved in the transaction with the 'manufacturer' who supplies the phone. The 'stock manager' contacts the 'parcel services', who will be the executor, sending the mobile phone to the 'client'.

The 'client' may experience a problem with his mobile phone. He will then contact Telco. The person dealing with the call, must ensure that the 'client' receives a replacement mobile phone. This has been outsourced to TOP. It must therefore be arranged that TOP is informed of this and that the phone is sent from stock by the 'parcel service', who will then deliver it to the 'client'.

If the 'client' wants to terminate his mobile subscription, he will request Telco to do so. The 'subscription stopper' receiving this request, will request the 'client' to return his mobile phone (as the cell phone is only on loan). That phone must be returned to TOP's stock. The 'subscription stopper 'therefore *requests* TOP to accept the phone back in its stock.

7.2 Case 1: Cooperation Within a Department

7.2.1 Brief Description of the Situation

For an internationally operating bank, ICT systems are developed, maintained and operated for both business- and private services. There is a department in the bank that issues the assignments to the ICT department, checks whether the right functionality is delivered during development, and is responsible for accepting the end product. This department is referred to as 'Business' in this Case. The Business will in this situation grant assignments for the development of software for Customer Relationship Management, in ICT these are usually referred to as CRM applications. These are software systems that enable the sale of bank products, like entering into a loan or a mortgage, via the internet.

For this Case all you need to know is that there are two departments within the ICT section that need to work together in order to jointly realize the requirements of the Business. These departments are Architecture and Development. Architecture determines the framework to which systems, delivered by projects, must comply. These involve, for example, rules for the connections with other systems, but also for the technology that is to be used. Development builds information systems that meet the requirements of the Business and comply to the framework determined by Architecture. The development organization is set up in programs and projects. Projects delivery partial solutions within a program, they are jointly responsible for the delivery of the final result. A program manager manages the program.

As the technology used is reasonably new to this bank, the architectural framework has not been definitely determined, it is still being developed, just like the systems. Regularly new policy is formulated, which the developers must take into account. They experience this as a burden. They are already very busy with developing systems. And now they are also required to take into account the continually changing framework, which unfortunately must be immediately complied to. This frequently conflicts with the pressure to deliver new systems for the Business in as short a period as possible.

Architecture directs, among other things, the development of reusable components. Reusable components are software parts—for example a tested, working payment module—that can be reused in other systems, and therefore don't need to be developed again and again. They form the building blocks of a system as it were. In the construction industry reusable components are also being used, for example the concrete blocks with slots, allowing them to slot easily in one another, and reusable wall systems. This way of developing offers many benefits. It allows one to develop faster; a part of the system is after all already completed and tested. It is also easier to maintain the system. If a change is made in the component, that change is directly introduced in all the systems that use that component.

The architectural framework is still being developed, and not every project is given the time to process the changes. This leads to a situation with different versions of the components being used at the same time. The major benefit of reusing components is therefore lost. The version of the component someone needs may not yet be available, thereby delaying development. And a change in a component must also be processed in all the versions, not just a single version. Hence, the benefit of simpler maintenance is also lost. That is why management has given Architecture the authority to explicitly provide the developers with the assignment to stick to an agreed version of a component.

However, the developers are most dissatisfied as they have lost their freedom to choose. On the one hand because it is often not possible to develop and use the components within the tight time schedule of the project. On the other hand they are given very strict instructions on how the components must be built, while they consider their own creativity an essential part of their profession. In their eyes, the following rule applies: architecture describes 'what' and 'developers' determine 'how'. The latter is now taken away from the developers. The developers respond by 'pushing' back. They start to tell the architects how to do their job. This causes the tension to rise to such an extent that they cannot work together anymore. An unworkable situation arises—the architects and the developers want nothing to do with each other anymore.

The program manager realizes that something needs to be done to resolve this unworkable situation. All sorts of things have been tried: a 'softer' approach, such as guidance and coaching has been tried and had no result. People have been sent away, but the problem persists. A complex problem, particularly as the culture of this bank does not allow for directive management. 'Must' is not done. Nobody is able to get a grip on this problem. Time for Enterprise Engineering.

7.2.2 Question and Approach

The program manager asked himself the question: can I improve the cooperation between Architecture and Development all along the line? In an initial phase, meetings were held to find out what the problem was. The accusation most often heard on both sides is that 'they do things they are not responsible for'. The word 'ownership' also seems to lead to a lot of discussions. Each party considered themselves the owner of a component, which means as much as 'the one that is allowed to decide what happens to the components'. By the way, the term 'ownership' recurs in many more areas (processes, tools, systems et cetera.) and in all organizations. It always leads to (heated) discussions. Everyone feels that they are the owner, but there is always only one true owner.

In order to analyze the cooperation better, we decided to select a specific process where the problem of the lack of any form of cooperation was an issue. We selected the software components process for this purpose. This process, which had not been documented yet, should be a guide for both Architecture and Development for the way in which reusable software components were to be developed. In a number of joint sessions we were able to determine the actor roles and transactions involved in this process.

At the end of the day, an AS IS Construction Model was drawn up, showing the underlying construction of the software components process that was currently being used by architects and developers. It made clear that there were ten Elementary Construction Flaws (ECFs—see Chap. 5) in respect of this minor area of cooperation between Architecture and Development, these ECFs were responsible for the root causes of the problems in the cooperation.

7.2.3 The AS IS Model

In Fig. 7.2 we see the customer 'Business' (CA44) which requests an actor role (A4) in the Architecture organization to manage software components (transaction T45). Business is also the owner of the component, because the owner is the one who pays. (In day to day life that is very logical, but businesses tend to be difficult about this.) We see the Architecture organization, a Development organization and a separate department (Configuration Management department.) which manages all the components within the bank as a whole (actor role A16). All the red elements in Fig. 7.2 are ECFs. I will discuss each of them.

7.2.3.1 Determining the Reusability of a Software Component

The transaction 'determine reusability software component' (T3) was very unclear. This was caused, first and foremost, because the term 'software component' had not been clearly defined. The same applied for the word 'reusable'. What is reusable and what is not? The solution appeared to be a strict, broadly accepted definition of the two words. This allowed the parties to discuss the reusability of software components without misunderstanding each other. It then needed to be determined who decided whether or not such a component was reusable: the architects or the developers? The analysis showed that the competences needed to be able to perform the role of 'determiner reusability SW component' (A2) lied with Architecture. That role is the executor in Fig. 7.2.

7.2.3.2 Acceptance of a New Component

Everybody accepted the fact that the developers needed to develop components. In Fig. 7.2, this is reflected as Transaction T11, with the executor being the elementary actor role 'developer SW component' (A12). However, the acceptance of whether a component met the criteria of Architecture (a component had to be reusable among other things) was done by the developers themselves. That is seen in Fig. 7.2 as a red line from transaction T11 to actor role A32.

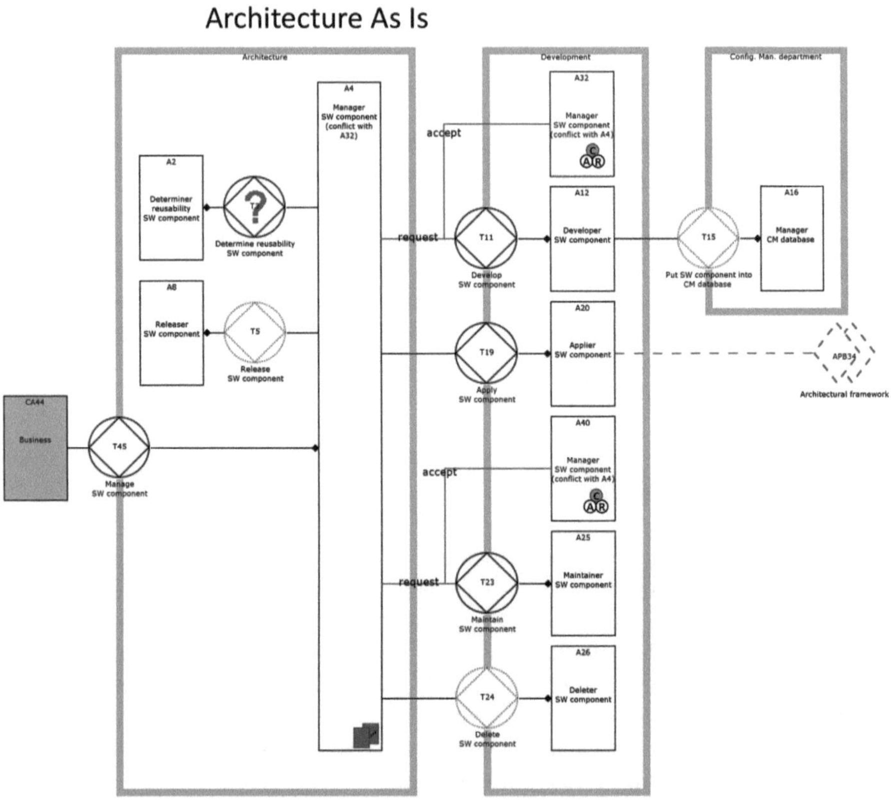

Fig. 7.2 AS IS construction model reusable software components with ECFs

In this case, there was no deliberately delegated responsibility by Architecture to Development. The developers incorrectly took it on themselves to bear this responsibility. Another problem that arose was that the developers responsible for accepting the new component did not have the right competences to do so. An architect is much more aware of the rules that the components must comply with, than a developer is, and he is also better able to apply these rules in practice. This is visualized in Fig. 7.2 with the red 'C' in actor role A32.

7.2.3.3 Acceptance of a Changed Component

Something similar happened for the performance of maintenance on the existing component; see in Fig. 7.2, transaction T23 and actor role A40. Again the developers accepted the changes of the components themselves, while Architecture should do this and is also better able to.

7.2.3.4 Release of a Component

Regularly the developers did not know that a (new version) of a component was available and could be used. This was a missing transaction; see T5 in Fig. 7.2. It had been discussed internally at Architecture that a new component could be used, but Architecture failed to tell the developers, partly as a result of the poor cooperation.

7.2.3.5 Use of a Component

Every now and then things went wrong with how a component was used (see transaction T19 and actor role A20 (Applier SW component)). Sometimes a component needs to be used in one way, while in another situation it might be used in a different way. This has all been documented in the Architectural framework. The framework had been drawn up by Architecture in this particular case, but it was not used by the developers. This is visualized in Fig. 7.2 by information bank APB34 (Architectural framework).

7.2.3.6 Removing a Component

Sometimes a (version of a) component is not allowed to be used anymore. It is old and Architecture has good reasons not to use it anymore, for example because it no longer complies to the framework drawn up at higher levels in the organization. In this ICT organization such outdated versions were never removed. As a result sometimes developers used a (version of a) component that should never have been used anymore. Often, people found out later when there was no more time to restore the error.

7.2.3.7 Storing Component in a Database

It is determined at a higher level in the organization that all the components of the bank must be included in a comprehensive database. The architects felt that the developers should store the newly made or adjusted components in the database, and the developers felt that the architects were supposed to do this. This led to heated discussions. In this situation the opposite of what we saw earlier applied: nobody wants the responsibility.

It turned out that someone else in the organization was responsible, i.e. A16 ('manager CM database').

7.2.4 The TO BE-Model

Figure 7.3 shows the desired Construction Model, in which all the underlying ECFs are resolved.

7.2.4.1 Determining the Reusability of a Software Component

Once Architecture had given clear definitions for 'component' and 'reusability', the developers had no problem with the architects taking the decision on whether or not a component was reusable. Clarity on definitions often results in immediate result and change.

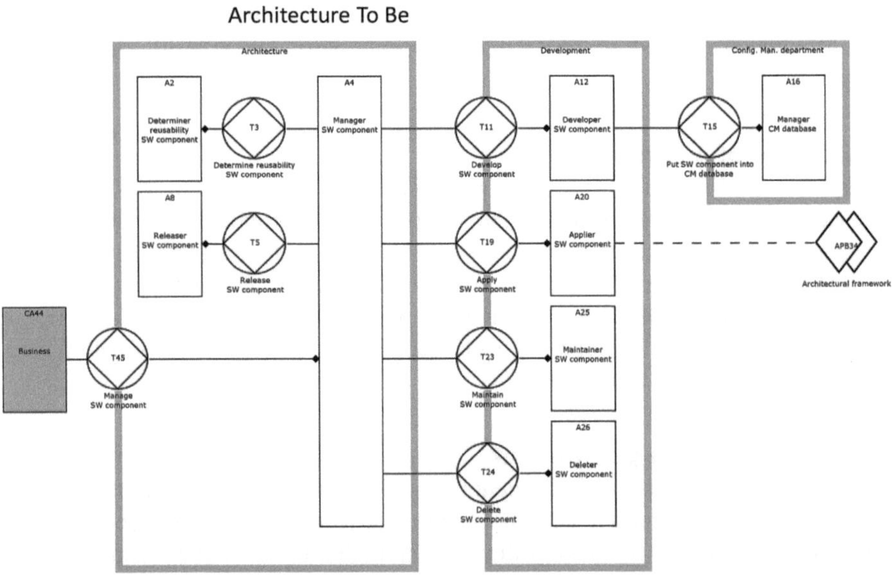

Fig. 7.3 TO BE construction model reusable software components

7.2.4.2 Acceptance of a New Component

Developers had not realized, also due to the time pressure they were under, that the architects had to formally accept a component. 'It was actually very logical', they said afterwards. Creating clarity in responsibilities and making them objective make people more willing to change.

7.2.4.3 Acceptance of a Changed Component

The same applies as for the acceptance of a new component. Again it was logical to have the architects accept the change in a component.

7.2.4.4 Release of a Component

This time, the architects had to say mea culpa. They hadn't realized that it might have been useful to inform the developers of a new (version of a) component. This was naturally also caused by poor cooperation.

7.2.4.5 Use of a Component

The architectural framework had been established, but had been imposed to rigidly. Architecture should determine the 'what', the developers the 'how'. But the framework also restricted the 'how'.

Once the architects and the developers were able to agree on the difference between 'what' and 'how', it became necessary to look into the architectural documents in detail and to remove anything that did not belong there. It was a painstaking process, but it did result in an Architectural framework that all the parties were able to accept. This framework was then applied for the use of a component. The framework was made available for everyone and explained during a number of sessions. The problem of the framework being ignored had been resolved.

7.2.4.6 Removing a Component

All the parties considered it no more than normal that unused components had to be removed. It was agreed that Architecture would give the developers the assignment thereto, should this ever occur.

7.2.4.7 Storing Component in a Database

Once it was clear that another party in the organization was responsible for storing a component in the comprehensive database, and that he was the only one who could perform this activity, it did not take long for the arrangements to be made with that party. Developers now send new or changed components to that party once they have been accepted by the architects, so that they can be stored in the database.

7.2.5 Results

The lack of cooperation had major consequences for the quality of the systems that were provided, particularly in respect of stability and maintainability. After the ECFs had been resolved and were included in the newly documented process—the software components process—all the parties started working according to these new arrangements. The cooperation became much better. A meeting that had been arranged to bring parties closer together and that often led to shouting and slamming of doors was no longer required within 3 weeks of working according to the new process, and was therefore no longer held. People like each other again and showed an interest in each other. This new cooperation also had an effect on other work processes. Within 2 months the poor cooperation was no longer visible. Without firing people, without being directive and without soft half-measures. The original problems never returned. Three years after implementing the changes, they have turned out to be permanent.

7.3 Case 2: Cooperation Between Departments

7.3.1 Brief Description of the Situation

In an internationally operating company an advice module is being set up for employees who maintain contact with the customers. The module is a tool which is supposed to provide support during the sale of corporate products. For this Case, we will split this company up into three areas of investigation: Business, Change and Continuity.

Business is responsible for formulating the requirements that must be met by the advice module. Business is the client for Change. Business is also responsible for accepting all the products that are supplied from Change.

The Change company is responsible for developing new ICT applications, such as the advisory module. Once the Change company has realized and delivered the new applications, the responsibility shifts to the Continuity company in order to keep these new applications running. The Change company is called Development from now on.

The Continuity company is responsible for operating all the applications. From now on we will call this department ICT Operations. ICT Operations is responsible for resolving any failures as soon as possible. Their highest priority is to make sure that all the applications continue to function as they are meant to do, and therefore that all applications continue to be available for employees and customers.

ICT Operations and Development must work closely together for the realization of the advice module, in particular during the 'production acceptance test' phase. The application has then already been tested to determine whether it does what it should do and how it responds in a chain of applications that are closely linked. The production acceptance test should determine how the application runs in the ICT Operations environment and whether this application is operable to an adequate degree. This ICT Operations environment has a strong resemblance to the ultimate environment in which the applications must run, i.e. the production environment. After a successful production acceptance test, the application can be taken into production and the employees and customers will be able to use it.

The project that must deliver this advice module is experiencing a lot of problems. For example, the cooperation between ICT Operations and Development is very bad. Both departments are particularly dissatisfied with each other and express the issues they run up against rather bluntly: they accuse each other of 'not being able and not willing'. It is also unclear which role exactly Business should adopt in this cooperation.

7.3.2 Question and Approach

The manager responsible for Development is quite unhappy with the bad cooperation with ICT Operations. Despite the several attempts that have been made to resolve this issue and which did not succeed, he asked if Enterprise Engineering could be used to solve this issue.

The cooperation that was looked at was the delivery phase when Development delivers a version of the advice module to ICT Operations which still needs to be tested. This delivery is accompanied with a complete package including documentation and code. During the delivery, it is checked whether everything is present and whether all the products meet the relevant requirements. If the delivery does not meet these requirements, it is sent back. Development should then prepare another delivery, after having resolved all the deviations. This process is continued until ICT Operations feels satisfied with the delivery.

7.3.3 The AS IS Model

In order to clarify the construction flaws in the cooperation between Development and ICT Operations during the activities performed to the advice module, the AS IS situation was first drawn up, in order to roughly clarify the cooperation between Business, Development and ICT Operations. In Fig. 7.4 this AS IS situation has been modeled, complete with Elementary Construction Flaws (ECFs—see Chap. 5).

On the left of Fig. 7.4 the Development organization has been drawn and ICT Operations is on the right. Business is not represented as it played no part in the delivery to ICT Operations. The 'project executor' (A148) requests the 'delivery acceptor' (A144) to accept the delivery, and the 'deliverer' (A149) requests the 'into production deployer' (A143) to take the advice module into production so that employees and customers can use it.

As Fig. 7.4 shows, ICT Operations has been assigned the actor role 'delivery acceptor' for the new advice module. The problem was however that this actor role did not have the authority to accept this delivery. The conditions that the delivery had to meet, the functional requirements, were after all not determined by ICT Operations, but by Business. Business should have been in control here: they made the request to develop a new version, and therefore they should also do the acceptance. Think of the universal pattern of cooperation: the one making the

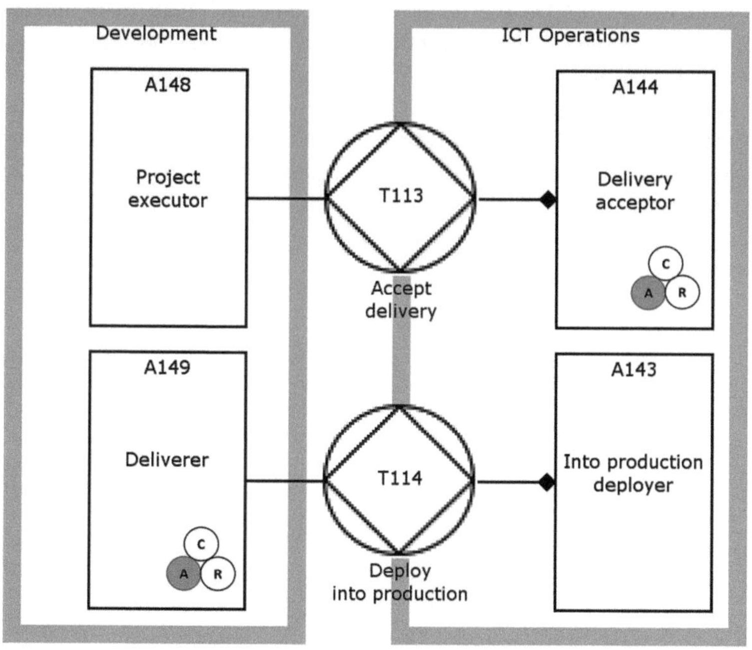

Fig. 7.4 AS IS construction model cooperation business, development and ICT operations with ECFs

request is also the one responsible for the acceptance, unless the latter responsibility has explicitly been delegated (but there was no mention of that in this situation). That is why the red 'A' has been placed in actor role A144, which means that actor role A144 does not have the authority to accept the new advice module. During the course of the years, ICT Operations had gradually taken the responsibility to accept a delivery.

In Fig. 7.4, we also see that the 'deliverer' (A149) requests that a certain release be taken into production. Again, this person is not authorized: the Development department shouldn't make this request, but Business should. Again the developers take a lot of the authority. And again there is no explicit delegation of responsibilities of the Business to Development. This is visualized in Fig. 7.4 by a red 'A' in actor role A149.

7.3.4 The TO BE Model

In the TO BE situation visualized in Fig. 7.5, these problems have been resolved and the roles have been moved to the departments that have the authority to perform those roles. The 'project executor' (A155) requests the 'product owner' (A149) to accept the delivery. In order to be able to take that decision, the product owner needs advice from ICT Operations: how well did the advice module pass through the production acceptance test? Does the new advice module not lead to an explosion of incidents? Do the expected incidents not result in an unacceptable increase in costs? The 'advisor production test' (A154) provides that advice. However, it is just that: an advice. The product owner determines the risks and accepts the advice, but ultimately takes his own decisions. After all, the interests of Business can exceed the risks involved with the advice module. Once accepted by the 'product owner', he will ask ICT Operations to take the advice module into production (transaction T144).

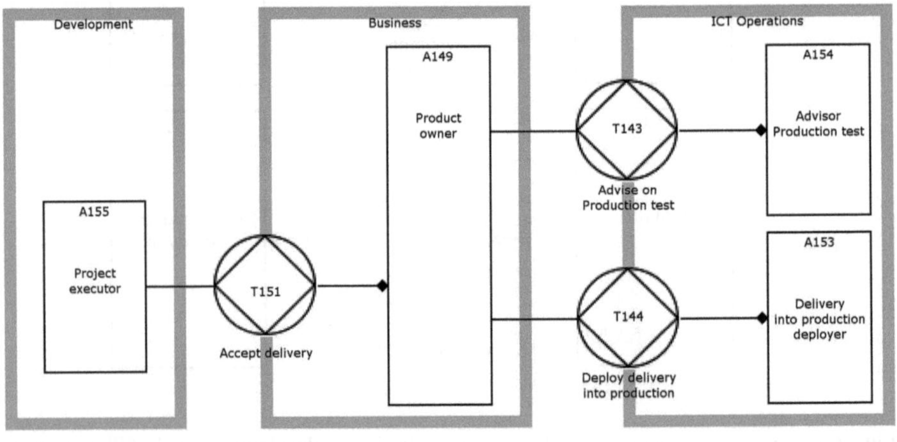

Fig. 7.5 TO BE construction model cooperation business, development and ICT operations

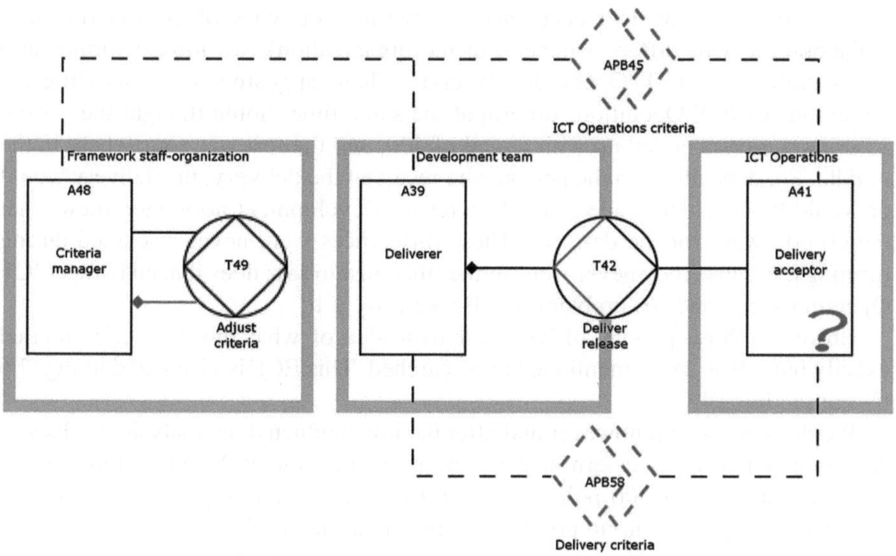

Fig. 7.6 AS IS construction model delivery from development to ICT operations with ECFs

After the high-level analysis has been drawn up, the responsibilities lie in the right departments and a detailed analysis of the cooperation between ICT Operations and Development was started. Soon it became clear that another party needed to be looked at. This was a department which managed and controlled all the rules and procedures that applied to the organization as a whole. We call this party in this case the 'framework staff organization'. The AS IS model appeared to have more than 120 Elementary Construction Flaws (ECFs). It would be too much to discuss them all in this book. But I will discuss the most important ones.

7.3.5 The AS IS Model

In Fig. 7.6 we see the framework staff organization maintaining the delivery criteria, the development team taking care of the delivery and the ICT Operations department accepting the delivery.

7.3.5.1 What Are Delivery Criteria?

Firstly there appeared to be two types of criteria the so-called 'delivery criteria' and the 'real' ICT Operations criteria. The delivery criteria (represented in Fig. 7.6 by databank APB58) were the criteria which the delivery needed to meet. These criteria were mainly administrative: did we receive everything we should have, and is everything in the right order? The ICT Operations criteria (represented in Fig. 7.6 by databank APB45) were the 'real' content-related criteria: which criteria need to be met by the advice module in order for it to be taken into production?

After this division had been made between both types of criteria (in close collaboration with different experts in the organization), we looked further into the definition of 'ICT Operations criteria'. There appeared to be six different definitions of ICT Operations criteria at the same time. Some thought there were 15 criteria, others ended up with a total of 400, and the rest was somewhere in the middle. So, depending on the person who received the delivery, the delivery would or would be considered adequate. As a result, Development no longer knew what the criteria were for the delivery. These differences were never discussed during meetings, because they never went further than just to say: does it need to meet ICT Operations criteria? No problem, I will take care of it!

However, those people all had their own idea of what those criteria entailed exactly but it was never mentioned or researched. This ECF is visualized in Fig. 7.6 by transaction T42.

We did not give up however and after having conducted an analysis we discovered that it had been determined at one point that the ICT Operations criteria consisted of a list of 300 points, but unfortunately only few people remembered this. This was too much, no project can meet that much criteria even if you really did your utmost. ICT Operations acknowledged as much. The solution suggested was to determine the relevant ICT Operations criteria for each separate delivery. An approach was adopted on which both parties agreed: a selection of criteria was made beforehand, and if criteria were added or removed along the way, ICT Operations would explicitly inform Development thereof. Having ICT Operations involved at such an early stage during the development of applications is fairly unique but has its advantages—provided the cooperation is smooth.

7.3.5.2 Who Determines the Generic ICT Operations Criteria?

The 'criteria manager' (A48) changed the ICT Operations criteria as he saw fit, but failed to inform the rest of the organization; a case of undesired self-activation (transaction T49). As it later turned out the list of 300 generic ICT Operations criteria were drawn up and maintained by a third party: the framework staff organization. Therefore the framework making organization is the 'owner' of this list, rather than ICT Operations. We were able to get rid of this misunderstanding in this way.

7.3.5.3 Who Accepts the Delivery?

As the two types of criteria were swapped, leading to Babel-like confusion, it also became unclear who actually was the acceptor of the delivery. This did not concern content-related acceptance of the advice module, but the administrative acceptance. This is visualized in Fig. 7.6 by the question mark in A41.

7.3.6 The TO BE Model

Figure 7.7 shows the desired Construction Model in which all the underlying ECFs have been resolved. I will discuss them one by one.

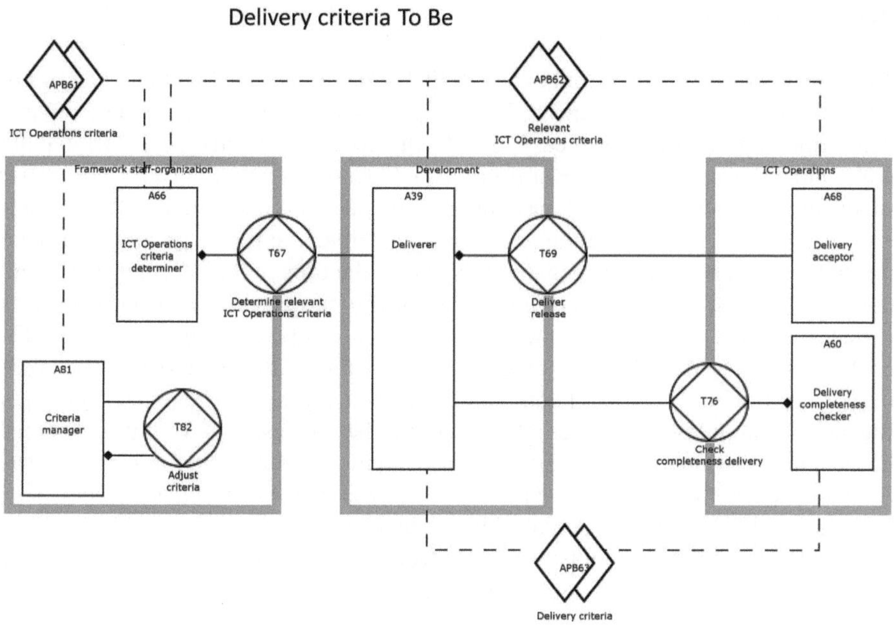

Fig. 7.7 TO BE construction model delivery from development to ICT operations

7.3.6.1 What Are the Delivery Criteria?

Development asks the framework staff organization (and not ICT Operations) what the relevant ICT Operations criteria are (T67). These relevant ICT Operations criteria are made available to all departments (APB62), so that everyone knows that these are the only relevant management criteria. This also prevents the possibility of a single criterion suddenly being added.

7.3.6.2 Who Determines the Generic ICT Operations Criteria?

The framework staff organization manages the generic ICT Operations criteria. They knew this already, but otherwise nobody was aware of this. What is new is that they will now inform the other departments of this (APB61), and that they will also do this if any change is made to the list of generic ICT Operations criteria.

7.3.6.3 Who Accepts the Delivery?

It is now clear that there is a difference between delivery criteria and ICT Operations criteria. The delivery criteria are assessed by the 'Delivery completeness checker', a role of ICT Operations (A60). This assessment only concerns the completeness of the delivery.

7.3.7 Results

Thirty ECFs were selected from over 120 ECFs found to be dealt with first and were given priority by the management. By having the underlying causes of the problems determined objectively, it was easy to make the necessary changes. The fact that the communications were far more uniform, ICT Operations was involved at an earlier stage and the introduction of new definitions such as relevant ICT Operations criteria, generic ICT Operations criteria and delivery criteria, greatly improved the cooperation between the departments during the production acceptance test and made it far more efficient. The timeframe for this process was reduced from 7 to 4 weeks, an efficiency improvement of over 40 %. And these were 'just' 30 ECFs.

The quality of the advice module is also very high. After the module was taken into production, the number of complaints reported appeared to be much lower than for similar modules. These complaints could also be resolved within the agreed norm time. And finally a delivery size measurement showed that much more has been delivered than the other processes realized before the changes had been implemented: productivity increased by more than 30 %.

Two years later, more ECFs were resolved and now the processing time of the production acceptance test has been reduced to 3 weeks. It still continues to improve.

7.4 Case 3: Purchasing Building Materials (VISI)

7.4.1 Brief Description of the Situation

All the cases in this book concern real situations, existing methods or existing processes that do not function as they are supposed to. The introduction of the process was already accompanied by the introduction of Elementary Construction Flaws (ECFs—see Chap. 5), and these have only grown in number during the course of time. In a situation such as this, a choice will need to be made: to take the existing situation and try to improve the ECFs, or to design a whole new work method and let go of the old method. Usually managers will opt for the first scenario. In this case they did not. A new work method for the cooperation was constructed from scratch. The case is situated in the construction industry.

In 1998, a number of parties in the construction industry decided to make clearly defined agreements on the mutual cooperation for the areas where parties worked together in construction projects. This cooperation was set up to allow parties to cooperate and communicate in a fast and flexible way, and would result in the parties being able to find each other much more easily. The construction industry requires many parties to actively work together (often dozens per project) but there are also varying ways of collaboration. If a building project is to be successful, a good division of responsibilities is crucial. Activities such as, for example, the rapid approval of changes, or making the latest version of the detailed drawing available on time, can have a huge effect on the quality, the runtime and the costs of a project.

The current technology offers an approachable support to the communication. However just the presence of this technology will not automatically resolve everything. As soon as more parties start communicating with each other, the chances of errors increase if no clear agreements are made on the communication process.

That plan from 1998 succeeded! In order to solve this problem, government, construction companies, engineers, branch organizations, knowledge institutes and ICT companies worked closely together to develop what is referred to in this sector as 'VISI'. VISI forms the foundation for digital cooperation. It consists of a set of agreements which can be applied to the construction process, regardless of the type of cooperation or contract used. This is referred to as the VISI method. The idea has been described and the content has been documented in such a way that it has resulted in an Open Standard that can be consistently applied, and that can support any form of cooperation and contract type and any outsourced task. This has all been achieved by applying what later became known as Enterprise Engineering.

In recent years, VISI has grown rapidly. It has been used in thousands of building projects since 2005, and every day new projects start using it. Rijkswaterstaat, ProRail, Dienst Vastgoed Defensie, Rijksgebouwendienst and Gemeentewerken Rotterdam have decided to include VISI in the requirements of all their U.A.V. '89-projects (specific type of projects; now referred to as UAV 2012). Many organizations even take it a step further, and apply VISI in all their projects. A VISI framework is defined for each project as a collection of roles, transactions, messages (elaborated information streams) and data, documented pursuant to the VISI system. That framework includes the agreements that must be adhered to in the cooperation. 'Standard' frameworks (such as that for the UAV) are used or reused, but each framework must be tailor made to suit the specific details of that project.

The VISI framework for a certain project therefore contains all the necessary elements and structure to deal with the communication agreements of that specific project in accordance with the prescribed patterns [1].

In this case we will discuss a generic VISI framework for the purchase of building materials.

7.4.2 The TO BE Model

Figure 7.8 shows the actor roles and transactions that play a role in the purchase of building materials. The following actor roles are involved.

7.4.2.1 Purchaser Building Materials (CA2)

The 'purchaser building materials' is involved in setting up contracts with the suppliers on behalf of the customer and in ordering (either on demand or not) of building materials from suppliers. Setting up contracts also includes making arrangements with regard to the commercial terms and conditions.

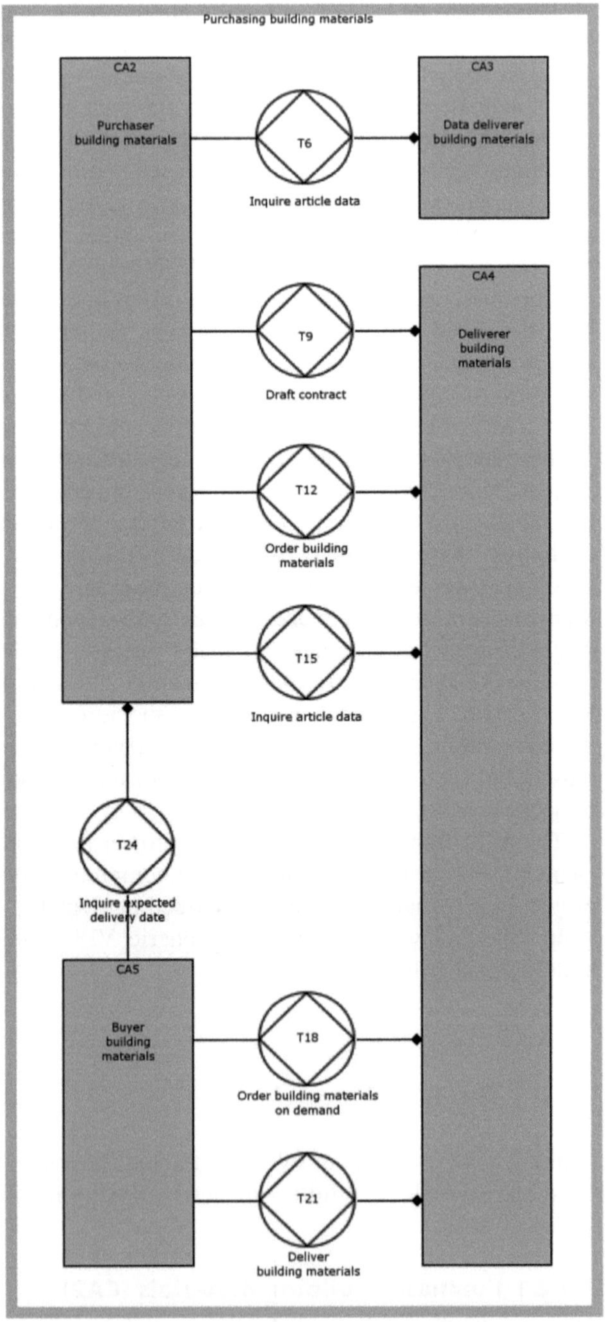

Fig. 7.8 Construction model purchasing building materials

7.4.2.2 Deliverer Building Materials (CA4)

The 'deliverer building materials' sets up contracts with customers and supplies the building materials ordered. He is not necessarily responsible for the manufacture, storage and transport of building materials.

7.4.2.3 Buyer Building Materials (CA5)

The 'buyer building materials' is the ultimate customer and orders the building materials on demand and receives them upon delivery. He is also responsible for the acceptance of the delivered building materials.

7.4.2.4 Data Deliverer Building Materials (CA3)

The 'data deliverer building materials' supplies article- and master data of the building materials to the customers on behalf of one or more suppliers.

7.4.3 Transactions Involved in the Purchase of Building Materials

7.4.3.1 Transaction T9: Draft Contract

This transaction involves setting up agreements on the conditions subject to which the trading parties purchase and deliver. The negotiating process on the conditions falls outside the scope, this is assumed to have taken place before this process. The 'purchaser building materials' will then sign a contract with the supplier ('deliverer building materials'). This is a framework contract with agreements on the prices of the building materials to be delivered. It is assumed that the commercial conditions for the contract have already been determined. The ordered building materials are eventually invoiced by the supplier.

7.4.3.2 Transaction T12 and T18: Ordering Building Materials

As the 'purchaser building materials' can come to an agreement without providing information on the commercial terms and conditions of that order, this order must be done separately, thereby referring to the concluded contract. It is possible that this is a single order, or an on demand order that forms part of a series of on demand orders under the same contract.

7.4.3.3 Transaction T21: Deliver Building Materials

This transaction is subject to the terms and conditions that apply to the delivery, receipt and transfer of ownership of the materials between the supplier and the customer. The customer, supplier and possibly logistic service provider must agree on a process for the delivery, receipt and inspection of the goods. The delivery takes place between the supplier and the customer. In practice, the delivery of building materials usually takes place to the direct customer/end user, and not necessarily to the purchaser.

7.4.3.4 Transaction T6 and T15: Inquire Article Data

When placing orders, the article data must be available. Whenever more data appears to be necessary in addition to the normal exchange of information, it is possible for one to request article data if and when necessary.

7.4.3.5 Transaction T24: Inquire Expected Delivery Data

Once the 'purchaser building materials' has placed an order with the 'deliverer building materials', he will notify this internally to the 'buyer building materials'. The goal is to notify which deliveries are to be expected. After the 'deliverer building materials' has sent the 'buyer building materials' a notification of the items sent, the buyer will inform the purchaser internally. This is done once the delivery has been accepted.

7.4.4 Conclusions

VISI has been successfully used in thousands of construction projects in the Netherlands and the system has now also been included in the ISO (ISO *standard* 29481, *Part 1* and *Part 2*). There is a growing international interest in the application of VISI. The method is widely spread all over the world. A wonderful example of Dutch innovation, with Enterprise Engineering (and the DEMO©—modeling technique) at its foundation.

The parties which have used VISI are experiencing the following benefits:

- The customer and the contractor in a project work together on one transparent digital communication file (for that specific project)
- It is always clear who does or has decided what at which moment, which means procedures are less time consuming
- Parties always have access to the most recent and complete project files, therefore reports are readily available
- Parties are able to enter into (new) collaborations fast and flexibly, thereby avoiding double work
- Clarity on the set up of a project within a short period of time
- Unambiguous communication between construction partners with VISI reduces the chances of miscommunication, construction errors and double work, and helps control the risks
- The application of VISI increases the transparency and traceability of the construction process
- The application of VISI has benefits in terms of the quality, efficiency, run time, costs and process control of construction projects. For example, the complexity in the cooperation model (the modeling pursuant to DEMO©) decreased by 95 %, filing costs can be reduced by as much as 80 % and the cost of failure (on average 7 % per project) can be reduced by a quarter.

7.5 Case 4: Complexity Control

7.5.1 Brief Description of the Situation

Consumers purchase and do more and more business via the internet, the mobile phone, iPad and other means and assume that these means can be used interchangeably and seamlessly.

This also applies to the customers of a large financial service provider. The Functional Maintenance department of this service provider acts as a link between the customers and the ICT organization and fulfils the client role for the maintenance of the ICT systems. Each day reports are registered of customers who fail to (correctly) perform their business via the normal channels. This gives Functional Maintenance a good insight into how the ICT systems work and allows it to perform its client role perfectly.

More and more often Functional Maintenance is faced with some annoying, structural problems. Realizing small changes in an ICT system costs relatively too much effort and time. The customers are also confused by an identical functionality (for example the approval of a financing request) having a different appearance in different applications. Also not all the related functionality has been implanted in one single application, which means that different applications must be used at the same time. This diversity and diffusivity is referred to as the 'complexity of the application landscape'. An application landscape consists of all the ICT systems and applications that are or must be interconnected one way or another.

7.5.2 Question and Approach

The questions that Functional Maintenance asks itself are: what are the real causes for the arising of complexity in the application landscape, how do we substantiate this with facts and what should happen to prevent that the complexity will only increase in future?

Many companies struggle with these questions and many programs have been introduced, trying to counter the increasing complexity in application landscapes. These programs are hardly ever successful. The exception is a large government organization that managed to phase out 49 % of its applications and drastically reduce the complexity by using Enterprise Engineering.

The manager of Functional Maintenance had heard of this case and wanted to see whether using Enterprise Engineering could lead to such results in his company. As the service provider had conducted various attempts, which all failed, and the word 'complexity' became increasingly laden, the ambition became not to reduce complexity, but to discover the root causes of complexity and to do something about that. 'Turning off of the tap' is a wise beginning if you wish to drain all the bath water.

After a significant number of interviews and sessions, the Construction Model of the Governance dimension was drawn up, including all Elementary Construction

Flaws (ECFs—see Chap. 5). The reason for such was that it was fairly certain that the existence of complexity is due to managers and not to people performing the work on the work floor.

7.5.3 The AS IS Model

In total more than 60 ECFs were found, that were part of the cause of complexity in the application landscape. As there are too many to include in this book, I have made a selection. Figure 7.9 visualizes part of the problems. There is not much to draw here, other than the customer and the contractor (the ICT department). On the square millimeter this model almost only consists of ECFs.

7.5.3.1 No Check on Complexity Increase

During the early stages of the development of new applications various stakeholders determine the requirements that the applications must meet. For example, those who will eventually work with the application have provided requirements regarding 'the look and feel' as well as regarding how the application should work. The system architects will then make the necessary technical choices to determine how the application will meet these 'look and feel' and functional requirements. The architects receive this assignment from Business. The requirements and the technical choices are jointly referred to as a 'global solution direction'. That direction is binding for all parties involved in the development of the new application: nobody is allowed to deviate from this anymore. However, no check for complexity is ever performed (transaction T4). Actor role A7 who performs the complexity check does exist, but is often so busy with his other main tasks, that he has insufficient time to do a complexity check. As the necessary

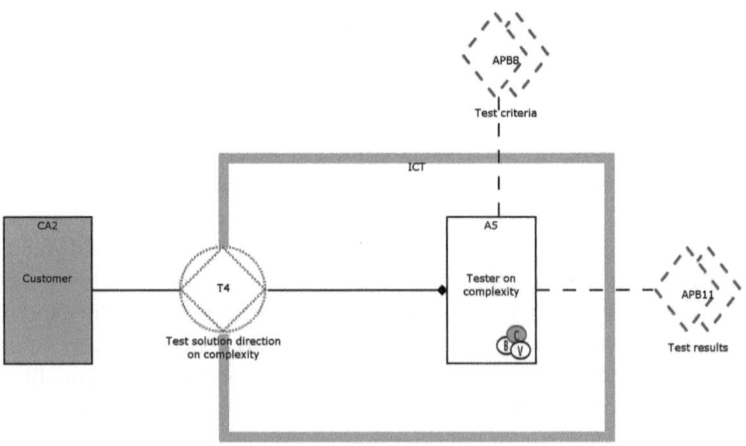

Fig. 7.9 AS IS construction model check on complexity increase with ECFs

check and feedback is missing in practice during the development process, applications are developed that generate a lot of technical complexity, for example because they are linked to other applications in an unnecessarily complicated way. These types of risks of additional complexity are not detected beforehand.

7.5.3.2 How to Check for Increased Complexity?

Another problem is the lack of objective check criteria (APB8). If those responsible for reviewing the prototypes even get around to doing this (which, as I have said before, is often not the case), they also only apply their own, independently devised, criteria which the application must meet. There are no generally applicable criteria that indicate exactly where the focus needs to lie. This subjective attitude leads to differences in focus and an inconsistent approach.

7.5.3.3 Documenting Check Results

It turns out that check results are also not documented (APB5). It is important to document the results, as such creates transparency. The institution ultimately approving the realization plan of an (amendment to an) ICT system, i.e. the customer, then would know what the money is spent on and what the risks and consequences are. As this service provider fails to perform checks during the development of the applications, the changes in applications or new applications are approved without knowing the consequences thereof.

7.5.3.4 Focus on Increased Complexity

Another important cause of complexity appears to lie in the focus of the customer, referred to in Fig. 7.10 as Business.

The Architecture department is responsible for making the technical choices, including the complexity of the technical solutions chosen. When Business requests the design of new applications from Architecture, Business mainly formulates criteria with regard to innovation and compliance to the law and regulations. This

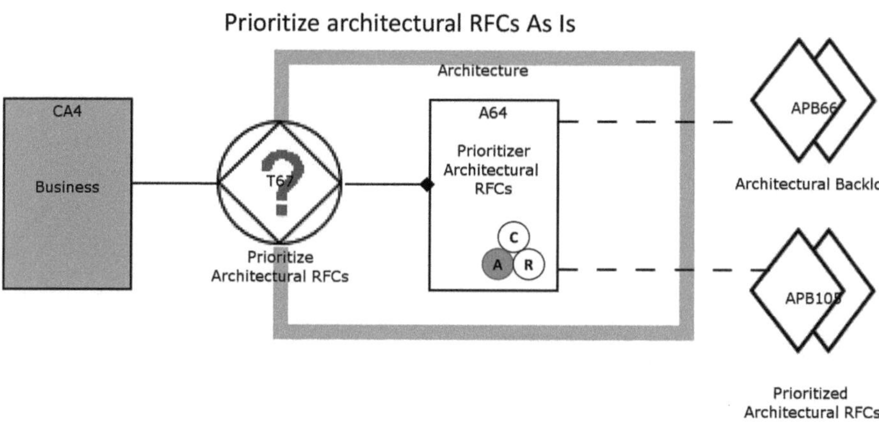

Fig. 7.10 AS IS construction model the role of architecture with regard to complexity with ECFs

leaves little room in the development budget for realizing the technical criteria, which is in Architecture's interest, for example for the prevention of any unnecessary complexity in the application landscape. We refer to this limited budget as the lack of authority (see actor role A64). Architecture has no budget to implement its own wishes. This prevents Architecture from being able to realize the responsibility for the reduction of complexity. The wishes of Architecture to introduce changes with regard to complexity, are included in Fig. 7.10 as Architecture RFCs (Requests for Change).

7.5.3.5 Focus on Complexity Reduction

Business should be requesting Architecture to provide insight in the prioritized proposals, the RFCs. Although not all requests for change concern the reduction of complexity, there are a few who do. This will cause more attention (and budget) to be given to changes that will allow the complexity to be reduced. And it mainly depends on priorities which changes are realized when.

7.5.4 The TO BE Model

In Fig. 7.11 all the ECFs have been resolved: the performance of a complexity check has been given to specific officers within the ICT department. Management gives them the time to structurally conduct the check for each change made to an application and for new applications. The check criteria for complexity have been drawn up by Architecture and are applied during the check. The check results are

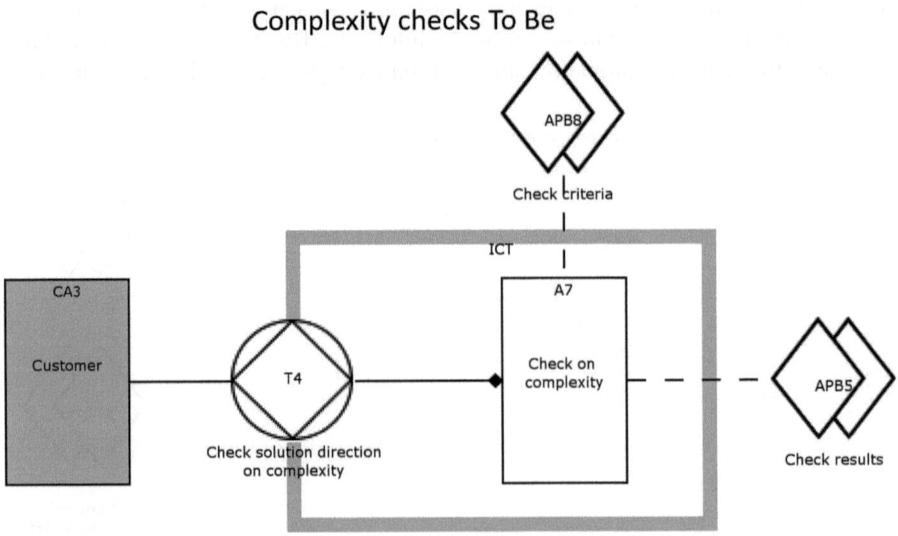

Fig. 7.11 TO BE construction model check on complexity increase

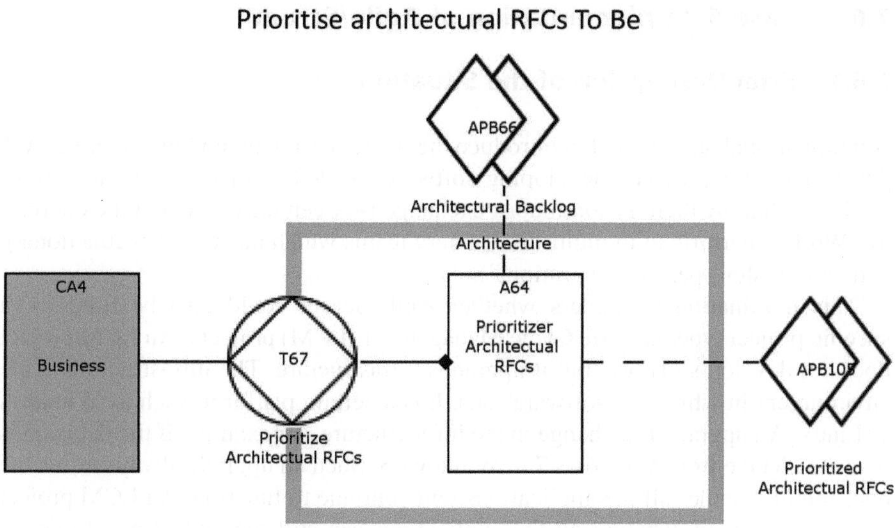

Fig. 7.12 TO BE construction model the role of architecture with regard to complexity

made available centrally, allowing everyone to be informed and Business to take the correct decisions on solution directions.

In Fig. 7.12 Business and Architecture interact with regard to the prioritization of architectural changes. All the wishes of Architecture are listed in a backlog (APB66), and after Business has given its priority to these changes, they are made available via a information bank (APB105).

7.5.5 Results

The introduction of new actor roles and new, clear transactions has made sure that a lot more attention is paid to complexity. It has become a lot more controllable. The number of complaints from the customer has significantly reduced. The renewed organization has allowed for complexity checks to be performed on the basis of fixed, objective criteria. Architecture, the department that has a lot of insight in technical complexity, has been given the opportunity to prioritize RFCs, allowing budget to be made available to deal with (anticipated) complexity. The effectiveness can now be monitored as the check results are documented. As more and more attention is given to controlling complexity, management is considering dealing with the complexity reduction on a more structured basis.

7.6 Case 5: Implementation of Agile/Scrum

7.6.1 Brief Description of the Situation

A multinational has decided to introduce the Agile/Scrum method in one of the ICT departments for projects, developing software. Agile/Scrum is a specific way of working whereby there is continuous exchange between the client and the contractor. Work is performed in multi-disciplinary teams which must deliver functioning software in short periods ('iterations').

This multinational wonders whether Agile/Scrum could also be used for a specific project type, i.e. Life Cycle Management (LCM) projects. An LCM project does not develop software, but it upgrades infrastructure. The infrastructure is the environment in which the software runs. It is a certain platform, such as Windows or Linux. An upgrade is a change in the infrastructure, for example if the decision is made to change from Windows 7 to Windows 8. Such an upgrade always raises the question of whether all the applications will continue to function. An LCM project is set up to review whether all the applications can still run and, if not, to detect what causes this and what needs to be changed to still secure the desired operation. The introduction of these types of changes in applications resulting from infrastructure upgrades is referred to as a 'migration'. LCM projects are usually large-scale projects that require the 'migration' of many applications.

7.6.2 Question and Approach

The manager of the LCM projects wants to know whether Agile/Scrum can be applied well and wonders which additional conditions must be met in order to perform an LCM project with Agile/Scrum. And if the decision is taken to perform LCM projects using the Agile/Scrum method in future, what would the desired profile of the relevant roles in an LCM-Scrum environment then be? In other words: which responsibilities, authorities and competences should the relevant employees have or receive?

7.6.3 The AS IS Model

The Product Owner plays a key role in a Scrum project. The person adopting this role, represents (the interests of) the project in respect of all the stakeholders of the project, such as the users of the end result (those who will be working with it), customers, decision makers, budget holders etc. The Product Owner must have an extensive mandate (authority). He should also be sufficiently competent in respect of the functional operation of the applications and of ICT. That is why an organization starting to work with Scrum often finds it difficult to find someone who can take on this role. In order to determine whether the role of the Product Owner, as defined in the Agile/Scrum method, connects to the roles that already exist in the organization, the role of Product Owner at this multinational was analyzed in terms of actor roles and transactions. In other words: which actor roles does the Product Owner perform, which transactions does he execute and which does he initiate? As we have seen an actor role is a combination of responsibilities, authorities and competences, which we will abbreviate here by (r, a, c). When all the (r, a, c) of all the actor roles, a Product Owner needs to perform, are added up this results in a total overview of what a Product Owner should be able to do, should be allowed to do, and should be willing to do.

In this specific Case, it will suffice to look at the level of the composite actor role of the organization as this application involves a sum of (r, a, c)s. Just to recap: an elementary actor role performs only one activity (or takes one decision), and a composite actor role can perform various activities (or take decisions). A composite actor role is visualized in a Construction Model with a gray color, as opposed to an elementary actor role, which is colored white.

In this Case, the content of all other roles and transactions in the work environment of the Product Owner are less important; my aim is to show that the role of Product Owner is analyzed in order to determine whether it fits in the organization construction.

In Fig. 7.13 the Construction Model of the Product Owner shows which transactions the Product Owner would initiate and which transactions he would need to execute himself for LCM projects, should the decision be made to introduce Agile/Scrum. Figure 7.13 also shows the composite actor roles which the Product Owner could encounter after the introduction.

The customer gives the Product Owner the assignment to conduct the migration of all the applications that need to be changed as a result of the infrastructure upgrade (see T5). To this end the requirements must be drawn up for each application that needs to be changed (see T6); changes need to be implemented in all applications (see T24); the governance team must be asked whether the realized changes are good enough (see T22); the applications must be deployed into production (see T7); and the implementation of each application must be approved (T17). And finally the old version of a migrated application must be removed from the production environment (see T23). After all, it will only be confusing if users of systems can/have to use two different versions of the same application at the same time. This may, for example, cause loss of data which is entered in an old version of an application by accident.

Next the responsibilities, authorities and competences the Product Owner must have in respect of the LCM projects were analyzed. There appeared to be many, as the group of stakeholders of an LCM project includes a lot of disciplines. These often involved dozens, sometimes even hundreds of applications, each having some dozen of stakeholders. This imposes additional requirements to the competences the intended (role of) Product Owner would need to have, for example as the 'ambassador' of the project in respect of all those different stakeholders: he must create a consensus, continuously collect information and mutually share information et cetera. The great diversity of stakeholders in an LCM project also makes it very difficult to organize a mandate (authority) for the possible role of the Product

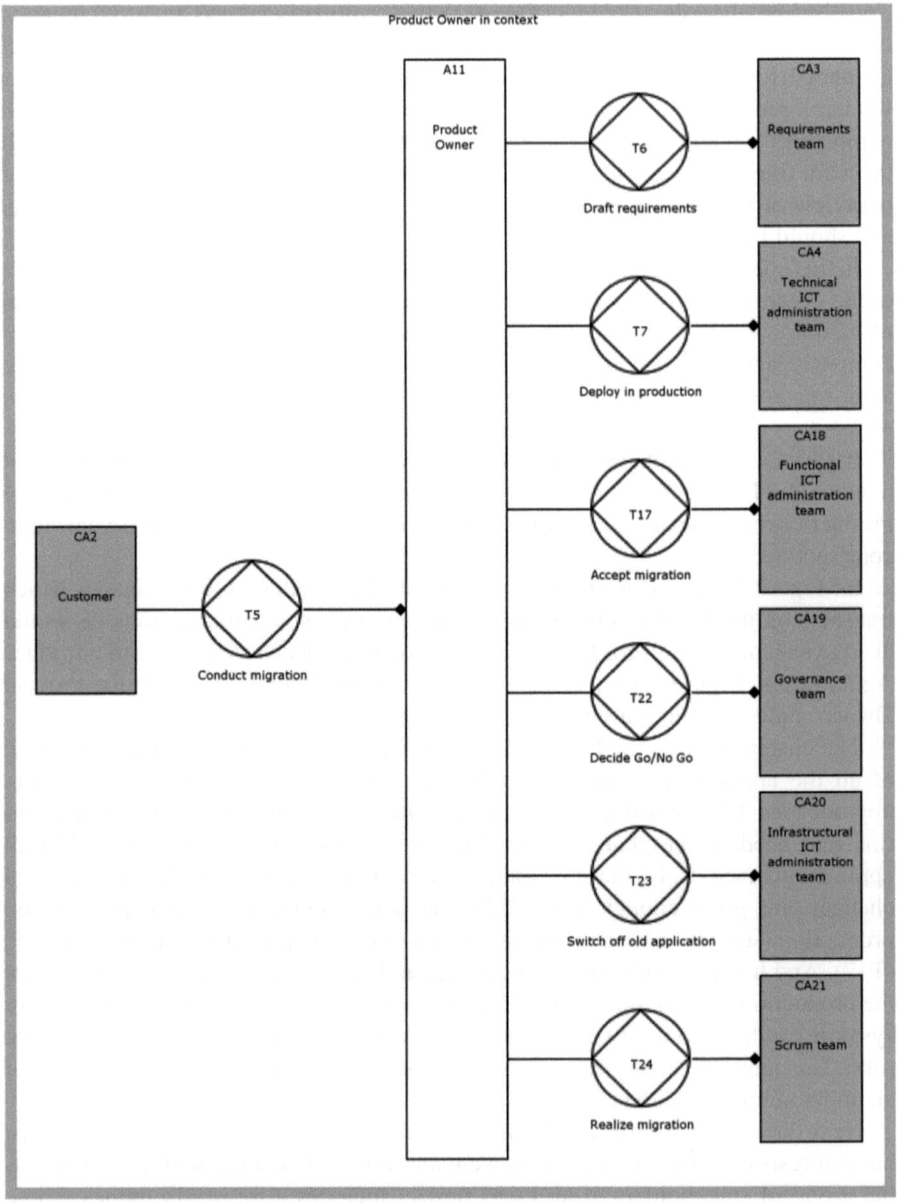

Fig. 7.13 Construction model of the product owner in LCM projects

Owner. This combination of competences and mandate indicates that the role should be performed by someone with great responsibility, therefore someone holding a higher management position. Also, a thorough knowledge of and insight into the technical aspects is required.

All in all, the question arises whether someone in this organization is able to meet these high profile requirements. An analysis of the current employees showed that this was not the case. The Scrum approach is characterized by a significant level of authority of the Scrum team. A Scrum team determines a lot for itself and should therefore be given a mandate by the organization. The analysis shows that such an approach does not suit the ICT department of this multinational. It is therefore very hard, if not impossible to apply Agile/Scrum in this organization and these types of projects.

A deeper analysis led to the conclusion that only part of the activities at LCM projects would be suitable for Agile/Scrum. This would still require a number of difficult conditions to be met: the role of the Product Owner must be split up further and a new role, a 'Project Coordinator', must be implemented next to the role of Product Owner.

The actor role 'project coordinator' is indicated in Fig. 7.14 in the context of an LCM project. He manages the non-Scrum teams as a classic manager (T13, T17, T21, and T25) and facilitates the Scrum teams (T29). A classic model is used for the managing the 'project coordinator': a program steering committee provides the assignment (T3). This steering committee also takes decisions in exceptional situations (T9). This differs from Agile/Scrum projects in which the Product Owner can take these decisions. In addition the decision to finish a project is also taken by the steering committee (T7). Without the 'project coordinator' as a spider in the web, it is not possible to responsibly implement Agile/Scrum for LCM projects.

7.6.4 Results

By applying Enterprise Engineering, we were able to show to this multinational that the investigated part of the organization in an LCM project did not have a solid basis for the role of Product Owner as required by the Agile/Scrum method. If the decision was taken to implement Agile/Scrum for LCM projects, a new role would have to be created. In view of the extremely complex way in which a possible new role would function, the final conclusion of the research was therefore that it was better not to apply Agile/Scrum for LCM projects.

These projects are still performed at the multinational in the old, project method and successfully so. This case only includes part of the full analysis. The total package of determined conditions for the application of Agile/Scrum in LCM projects has been abstracted to projects (therefore not only LCM projects). This package has been included in a corporate Scrum policy. The multinational has decided to review each new project that is to be started in order to determine

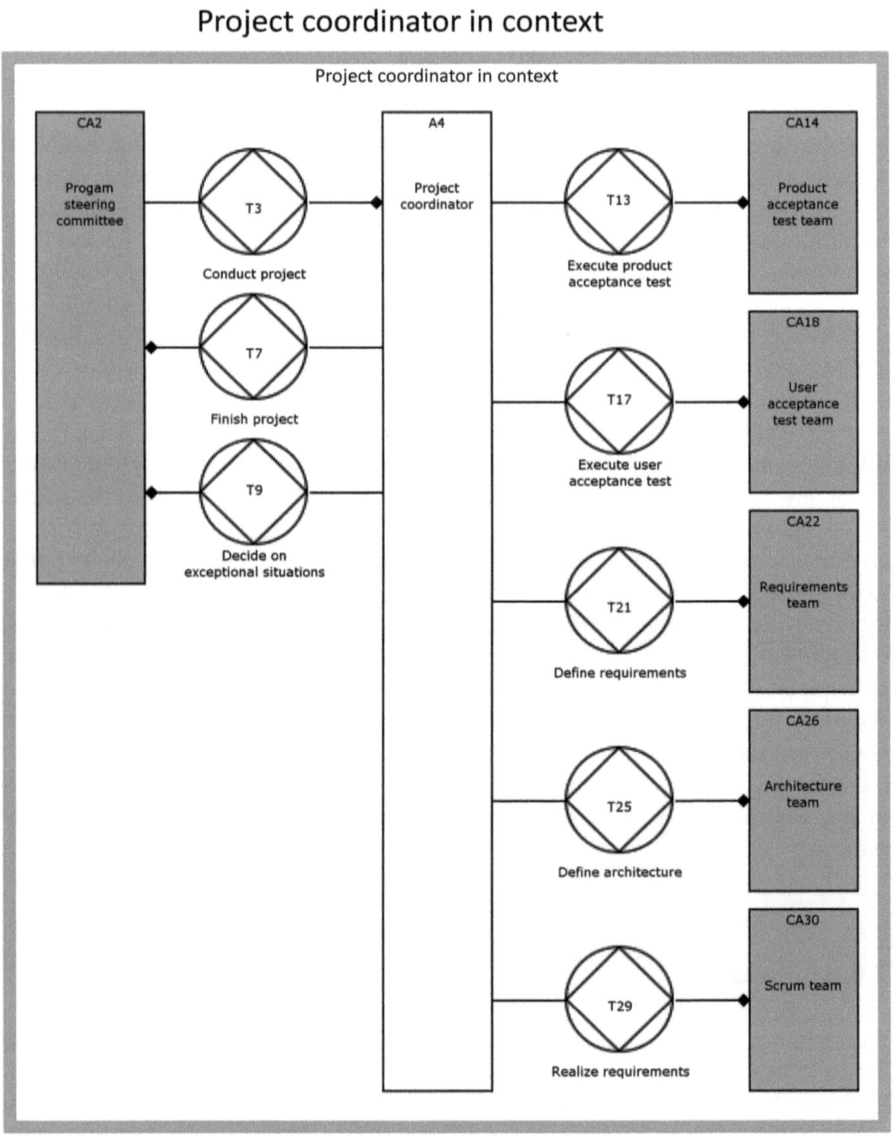

Fig. 7.14 Construction model of the project coordinator in LCM projects

whether the formulated conditions for a successful application of Agile/Scrum are met.

If this company had implemented Agile/Scrum for LCM projects without being aware of the consequences, this would without a doubt have led to a lot of problems. It is extremely valuable to have the kind of objective information beforehand which in this case was retrieved with Enterprise Engineering. Too many organizations often realize too late that the construction of their organization does not correspond with the responsibilities, authority and competences of the introduction of new roles required for a new method (in this case Agile/Scrum).

After a while, sometimes a couple of years, the new method will conflict to such an extent that resistance against it grows, and eventually the method is blamed for the fact that goals are not achieved. That is the end for this method and one will start looking for a new one…

7.7 Case 6: A Reorganization

7.7.1 Brief Description of the Situation

A government organization maintains and manages a large quantity of privacy-sensitive information for the central government. This organization manages all these data in a secure way and arranges the data supply to a number of clients, for example for policy making or scientific research. The activities of this organization consist of the receipt, including validation, storage and release of data. They explicitly do not draw conclusions from the data; this is done by the customer receiving the data.

The organization is still young and has seen a rapid growth within a short period of time. As often happens with organizations that grow rapidly and do not (yet) have the means to keep on hiring new employees, a lot of gaps need to be closed. Employees do have a position, but are required to do a lot of other work as well. Positions shift, employees are given other tasks continually, are asked to do things they are not good at, not everything is coordinated correctly et cetera. This causes the internal cooperation to run less smoothly than desired as well as confusion on the areas of responsibility. Who is responsible for what, and who do I need to contact for what? The organization has not taken the time to analyze, draw up and communicate their way of working in detail. This situation has been going on for a while now.

7.7.2 Question and Approach

The manager of the organization faces the challenge to turn this rapidly growing organization into an efficient and effective one. That is why the decision was taken to focus on the internal cooperation with the aid of Enterprise Engineering and to make proposals to improve the effectiveness. Effectiveness as in: what do we need to change or add in order to be better able to serve our customers?

7.7.3 The AS IS Model

This Case also involved a large number of Elementary Construction Flaws (ECFs). But I will again only discuss the most important ones.

Figure 7.15 represents a part of an AS IS Construction Model. We see actor role A2 'release manager' has been given a central position. In order to be able to manage privacy-sensitive data correctly, this organization has developed a system and a database for own use. As the central government and other customers regularly have new wishes, for example other cross sections through the data, new versions of the system and the database are regularly developed and taken into use. A *release* is an exact demarcation of the changes that are taken into use at a certain moment and consists of a collection of (various versions of) code files and accompanying documentation.

7.7.3.1 What Is in a Release and Who Determines That?

In the AS IS situation the release manager independently determines what is incorporated in the release and therefore which functionality will be taken into use. At any given time, the composition of the release can change, purely as a result of the situation (sudden wishes of customer, for example the option to request information via internet). This happens a couple of times a week. The developers simply cannot keep up with that many changes, let alone incorporate them in the programs. The 'release manager' does however have the authority to impose all these changes. This generates a lot of frustration, confusion and incomplete work of the employees. Almost every day work that had already been performed was skipped and new work had to be started.

7.7.3.2 When Does Which Functionality Need to Be Used?

Another problem is that there is no planning for the releases.

Employees therefore don't know what the deadline for certain functionality is, and are unable to prioritize their work. Regularly work is performed that is completely irrelevant in the short term. As soon as the manager is aware of this, he acts and wonders 'what people think they are doing'.

Not keeping a record of what a release entails, not knowing when the release needs to be delivered, is as a ship without a captain with loose freight: you have no idea when you will arrive where or whether you will deliver the correct freight.

7.7.4 The TO BE Model

7.7.4.1 What Is Included in a Release and Who Determines This?

The problem of the continual changes being made to the composition of a release is resolved by introducing a Change Board, in which various officers take part, including the 'release manager'; the Change Board recognizes and determines the impact of a change. Based on content-related information the Change Board can decide in consultation whether a certain change is included in a certain release, and

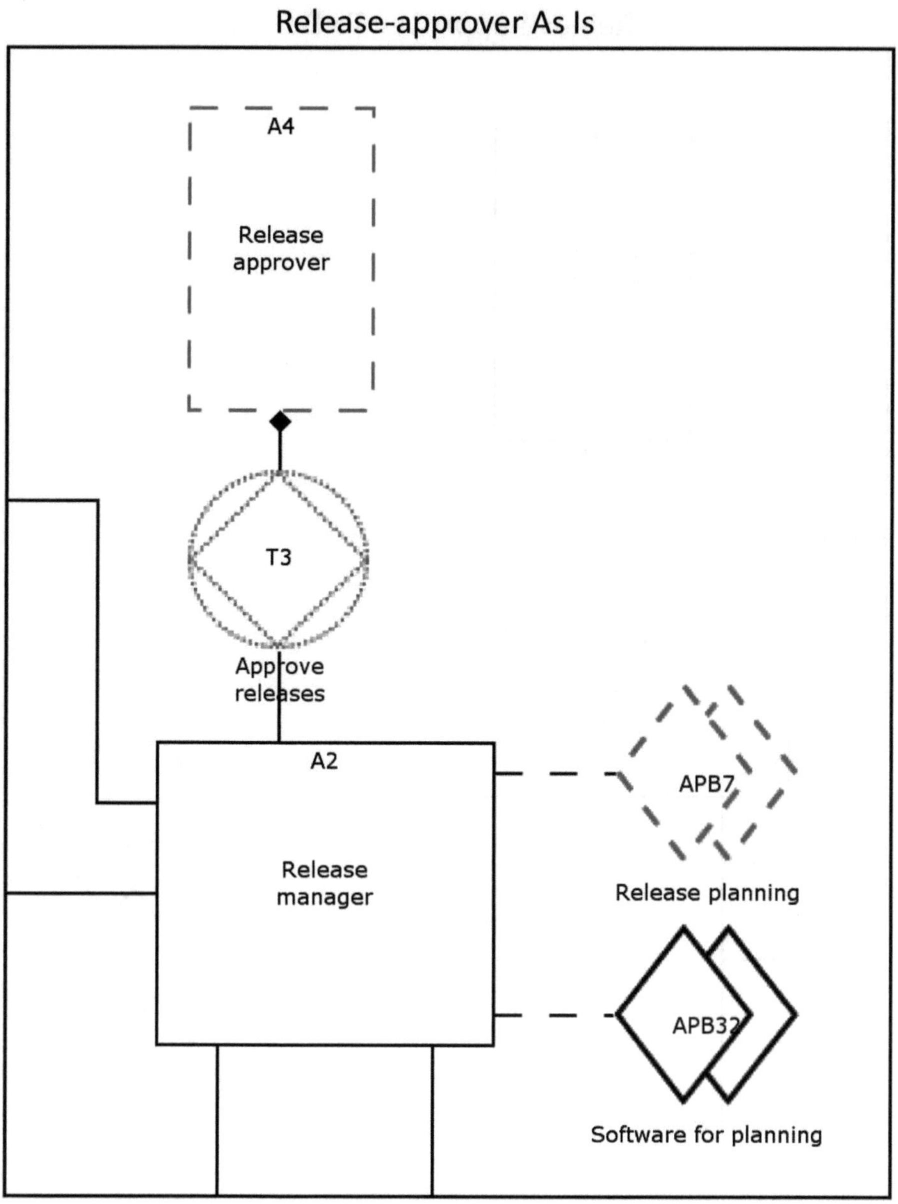

Fig. 7.15 AS IS construction model release management with ECFs

if so, in which release. The manager of the organization is the chairperson of this Change Board. In Fig. 7.16 this has been depicted by adding an actor role 'release approver' (A4) and a transaction 'approve releases' (T3).

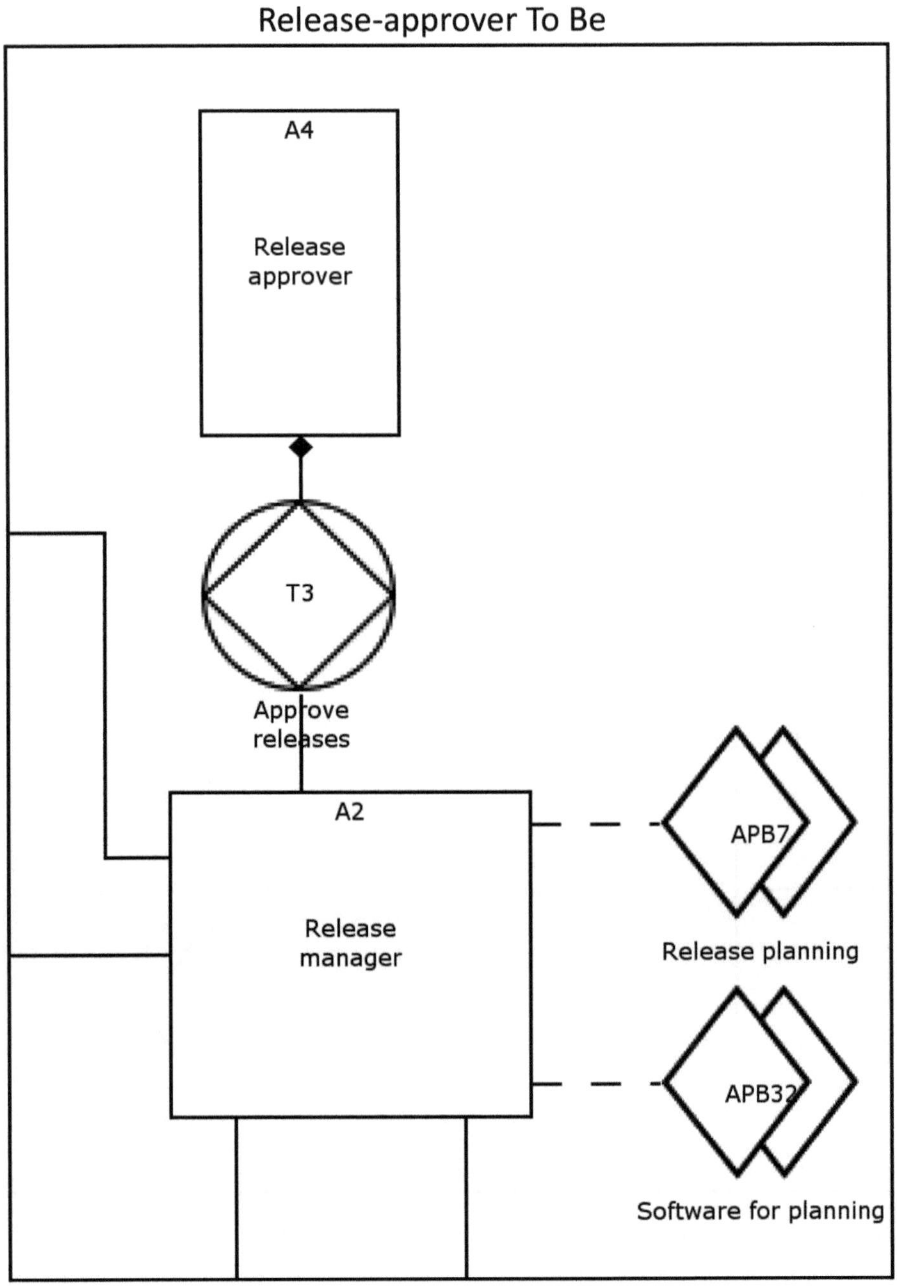

Fig. 7.16 TO BE construction model release management

Incident Solver As Is

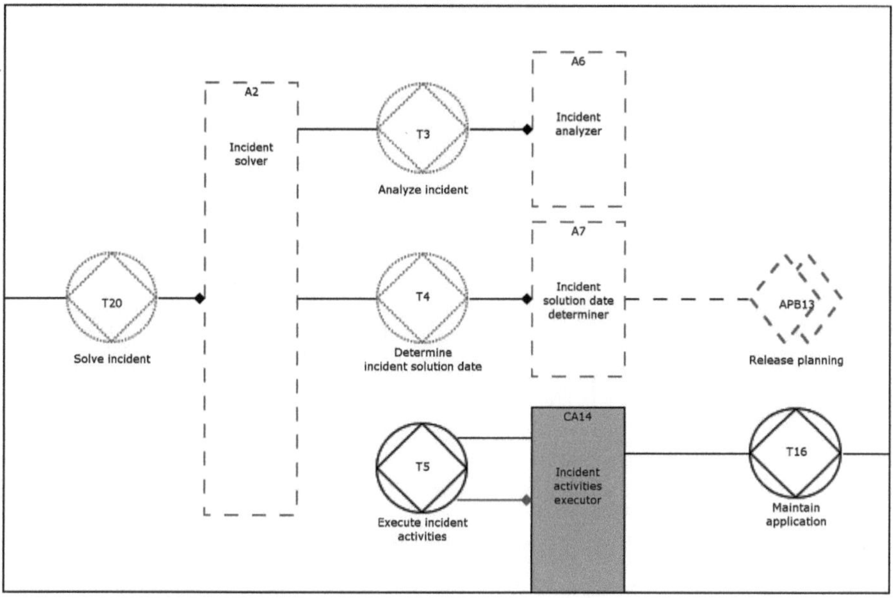

Fig. 7.17 AS IS construction model incident management with ECFs

7.7.4.2 When Do Which Functionalities Need to Be Taken into Use?

In order to ensure that the 'release manager' (A2) and all the developers stick to the planning, a release planning has been drawn up (see databank APB7 in Fig. 7.17), stating the content and the implementation date of each release. This release planning is maintained by the 'release manager' and discussed in the Change Board. Each release change, approved by the Change Board is immediately processed in a new release planning version by the 'release manager'.

7.7.5 The AS IS Model

Figure 7.18 shows a Construction Model of another part of the organization. This part of the organization resolves disruptions in the system and is responsible for supplying data to the customers of the organization. The customers report disruptions the moment they take place.

These reports are registered in an internal application as 'incidents'. It is the intention for these incidents to be resolved within an agreed time unit.

Figure 7.18 shows that the incidents are not being dealt with in the AS IS situation. Nobody has appointed anyone to analyze the incidents (transaction T3 and actor role A6), let alone that anyone is appointed to resolve the problem in question within a certain time (transaction T4 and actor role A7). The only thing that is done, is the registration of the incident (see CA14). The incidents are

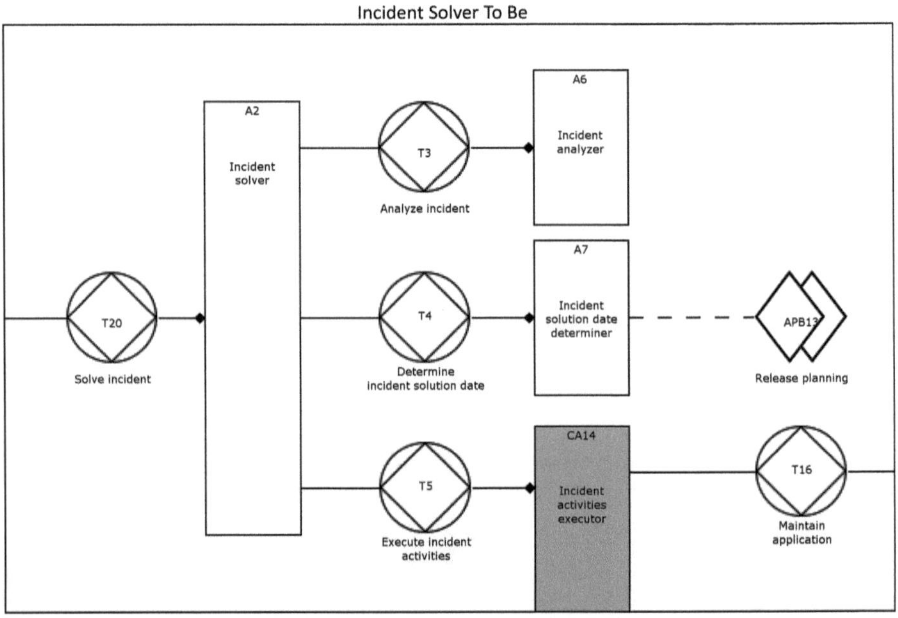

Fig. 7.18 TO BE construction model incident management

received centrally with the 'incident activities executor' (in Fig. 7.18 this actor role is CA14). But nobody is appointed to deal with those incidents. This 'incident activities executor' determines as he sees fit whether a certain incident must be resolved by the developers. But without any analysis, mind you. Analyzing and planning are never done. The 'incident activities executor' only passes reports on to developers when he considers it necessary, which is seldom. For the one suffering from the disruption and who filed the incident, it comes across as if nothing is done with the incident. In most cases this is true.

7.7.6 The TO BE Model

In order to improve this situation a number of new actor roles have been created. First of all, someone was made responsible for solving an incident (transaction T20). This is a crucial role, because without this role, nobody will feel the urgency to do something about an incident or to ensure that it is resolved within a reasonable time; in Fig. 7.19, this role has been added as actor role A2. This actor role is performed by the functional managers of the system.

An actor role has also been introduced who analyzes the incidents (A6). In case of disruptions, it is important to know whether it concerns a failure of the system or something else (a failure with the provider for example). If it does involve a failure

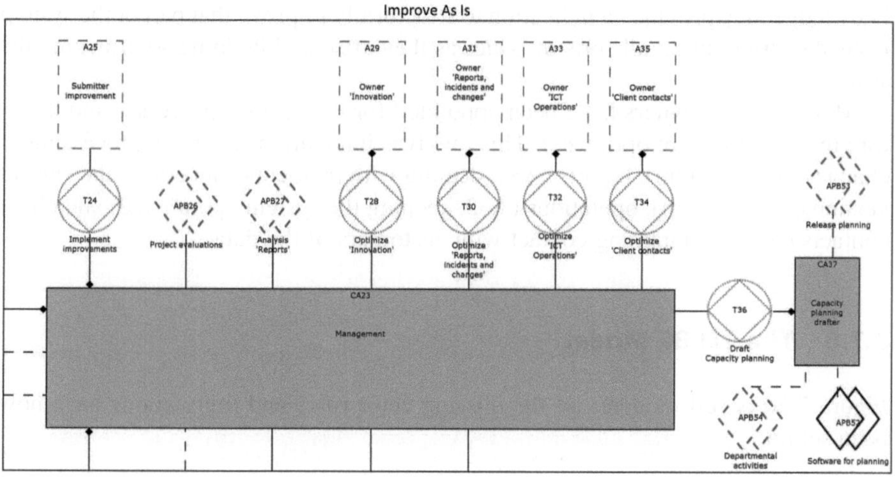

Fig. 7.19 AS IS construction model of the improvement organization with ECFs

in the system, then the question that needs to be answered is what the problem relates to exactly and how much time will it cost to resolve it.

Additionally, an actor role has been introduced who is responsible for determining the date when the incident must have been resolved (A7). The connection to the release planning is important in this respect (which had been absent as we saw before): the releases are subject to a tight schedule which does not allow for an incident to be resolved in the meantime. Customers will have to wait for the next release, provided it fits in that release (as briefly referred to above, a Change Board had been set up as a result of the investigation, who will determine whether or not something can be incorporated in a release).

Finally, the self-activating actor role (CA14) is turned into a normal actor role; it is not the intention for this actor role, performed by a staff employee, to take the initiative to instruct the developers to resolve a certain report. The person responsible for resolving the incident must report to this employee that work is allowed to be performed on resolving the incident.

7.7.7 The AS IS Model

Finally, in Fig. 7.19, we see another part of the organization, i.e. the AS IS Construction Model of the improvement organization. This figure makes clear which improvement opportunities the organization fails to use. It is for example not possible for anyone (A25 and T24) to submit improvements. No project evaluations are held (APB26) from which lessons can be drawn from the errors that were made. No analyses are made of the reports that have been received (APB27). A root cause analysis (determining the root cause of a problem) of the reports can expose which part of the system contains the most errors. Such knowledge may provide an indication to structurally improve that part of the system instead of firefighting all sorts of issues as they arise and in doing so drowning the system.

Moreover, no owners have been appointed for the various processes and therefore the system is not optimized. This involves innovation (A29, the processing of changes in the system), reports (A31, dealing with incidents concerning the current version of the system), operations (A33, keeping the system operational) and client contacts (A35, maintaining contact with customers of the data).

7.7.8 The TO BE Model

Figure 7.20 is self-evident: all the missing actor roles and transactions have now been set up.

7.7.9 Results

Thanks to the insight into the Elementary Construction Flaws (ECFs) this young and rapidly growing organization gained insight into the fundamental operation of the organization as it operated at the time. A Change Board was set up, which operates adequately. Employees are no longer surprised by sudden focus changes of the management. Everyone now has rest and insight into the organization due to the release planning being available to everyone. The customers of the organization are very happy about the changes that have been introduced in respect of the processing of the disruption incidents: something is done about it, people are also informed when the incident will be dealt with, and the incidents are actually resolved. The

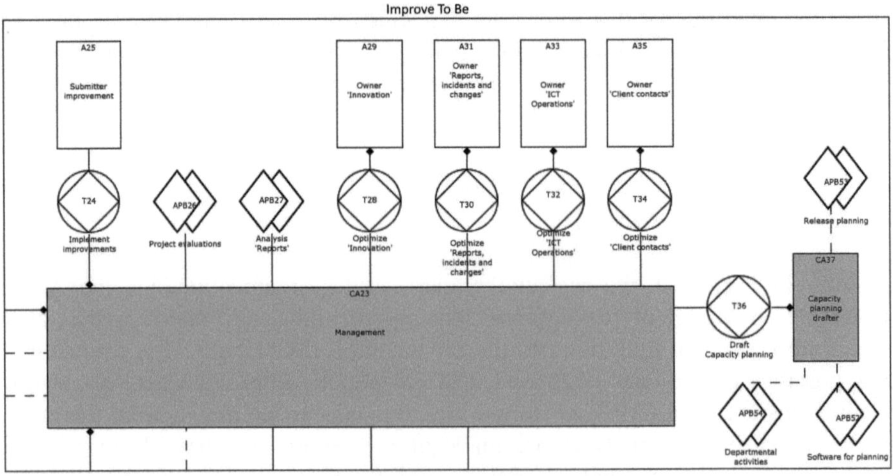

Fig. 7.20 TO BE construction model of the improvement organization

focus continually used to be on resolving the problems in the executing organization, but now the organization is ready to deal with structural improvement. On their way to becoming a self-learning organization.

7.8 Starting Small

Customers are always very impressed when they see how precise and specific Enterprise Engineering is able to pinpoint the problem. This is not so strange, as all the flaws in the organization were previously hidden in the black box. The discovery of such large number of Elementary Construction Flaws (ECFs), in particular, is always a real eye-opener. A full plate can make one greedy, but be warned: it is advisable not to tackle everything in one go. An organization would be best advised to start by implementing a new construction of a single process, and to then extend this to other processes. Most of the organizations that have applied Enterprise Engineering have started improvements in this way. The more one experiences the improved cooperation, the more organizations want to reconstruct it.

Reference

1. Braet, A. (Ed.). (2000). *Taalbeheersing als communicatiewetenschap. Een overzicht van theorievorming, onderzoek en toepassingen*. Bussum: Coutinho.

8 Enterprise Engineering in Respect of Other Themes

In Chap. 6 I have shown that there are many areas in which Enterprise Engineering can be applied. The Construction Model allows all the Elementary Construction Flaws (ECFs) in the organization to be restored (see Chap. 5 for the ECFs). The model also provides an excellent tool for takeovers, reorganizations, outsourcing processes and reduction of complexity. However, this does not mean that Enterprise Engineering can't be useful in other areas, which at first glance have little to do with the construction of the organization. We will see that the construction of the cooperation forms a solid excellent basis to deal with problems regarding corporate culture and HRM. It is even an excellent starting point from where to deal with political games and increase the satisfaction of employees. In this chapter, I will shed light on the link between Enterprise Engineering and various other subjects that play a role in organizations.

8.1 Enterprise Engineering and Corporate Culture

Cooperation consists of two aspects, if only one of those is not managed well, then the cooperation will, by definition, be ruined. It concerns questions '*what* shall we do?' and '*how* shall we do it?'

The 'what' of the cooperation is about 'what shall we do and what shall we deliver'. In other words: the 'what' concerns transactions in the construction of the organization. Judging by the Elementary Construction Flaws (ECFs) found in the Construction Model of organizations, the transactions on which the cooperation is based often contain errors. The fact that the organization still functions is mainly due to the willingness and efforts the employees.

The 'how' of the cooperation deals with the way in which transactions are executed. Because even in cases where the construction is 'healthy' it remains to be seen if everyone is satisfied by the performance. In other words, if not a single ECF is identified, the cooperation is in principle optimized, but only at the level of the construction.

T. Janssen, *Enterprise Engineering*, Management for Professionals,
DOI 10.1007/978-3-319-24172-2_8

The coordination activities request, promise, state and accept, can be performed in many different ways. Someone visits the baker and shouts ‘give me a roll and be quick about it’, which does make his acceptance criteria clear (he wishes to be served quickly) but the chances are that the baker will remove this customer from his shop. A baker who does not explicitly makes the promise (‘will be right with you Sir’), and will set to work, grumpy and silent, can count on strange looks. Also the state can be polite and formal (‘here you are, did you want anything else?’), but the baker can also decide to throw the bread from behind the counter to the customer as a jovial gesture. Finally, the acceptance may, for example be silent: the customer accepts the bread, pays and walks out the shop without a word. But he can just as easily conclude the transaction by being friendly and saying ‘thank you and goodbye’.

Enterprise Engineering does not prescribe *how* the coordination activities are to be performed—but it does prescribe the right people (actor roles) to perform the right transactions (for which they are responsible, authorized and competent) with the right cooperation partners (actor roles). The corporate culture can then be translated into a way in which transactions are performed.

An example. A bakery wants to have a reputation as ‘an honest and sincere baker’. The owner therefore feels it is important that when a *promise* is made to deliver a loaf of bread, the sales person thoroughly checks whether that bread is for sale. Strict conditions are applied. The bread must be baked perfectly golden and must be baked no longer than 2 h before. If the sales person sees that the bread is already too old, he will tell the customer that he cannot sell it. This is a referred to as a cancellation pattern. A bakery that is not so fussy about the sale of bread has exactly the same transaction in the Construction Model (‘selling bread’).

The transaction will take place more often than at the sincere baker. This is how the company culture plays a role in the construction of the organization.

The corporate culture also becomes visible in the delivered production facts. If an organization for example applies ‘a deal is a deal’ as core value, it is possible to focus on whether the production facts actually come about. If all the transactions are clear, then it can be exactly determined which promises are made and if they are in accordance with the states of the performing actor role (the executor). Obviously, the organization is free to determine how many ‘broken promises’ are acceptable.

Hence, Enterprise Engineering offers tools in respect of the corporate culture to objectively determine whether or not it operates in accordance with its own values. An additional constraint of applying it such will naturally be the availability of a Construction Model of the organization.

8.2 Enterprise Engineering and Human Resource Management

An organization has a certain corporate culture. This corporate culture determines to a large degree how the coordination activities are to be performed. It can also be agreed who monitors the corporate culture. This usually is a task for the HRM department. They may express that monitoring role for example in the following ways.

8.2.1 Exerting Control on the Way in Which a Request, Promise, State and Accept Are Performed

The HRM department could, for example, supervise by conducting audits on the coordination activities per actor role and this determine to what extent these are performed pursuant to the standards. Also, HRM could regularly offer employees courses in which they are taught how to perform the transactions in the right way.

8.2.2 Provide Clear Job Descriptions by Using Enterprise Engineering

As job descriptions are abstract and open to different interpretations, they hardly provide any support for the performance of the real work. They are also often outdated as they are not maintained properly. However, by using the Construction Model a lot of confusion can be avoided, confusions that tend to lead to ECFs (see Chap. 5). A job description may then list the exact actor roles the specific job is responsible for. This provides much more clarity than a vague description of tasks. Each actor role will clearly describe what it is responsible and authorized for. It also indicates the competences necessary for the performance of the actor role. A position is a sum of all the elementary actor roles that need to be performed by the position and therefore also needs an exact description for what this position is responsible and authorized. As well as the competences necessary for performing the position. If these job descriptions are checked by an expert, it can easily be established whether there is an overlap in positions or if gaps exist. In other words: are various people not unnecessarily performing the same work and are all the actor roles assigned? This will also allow all the unnecessary work to be skipped—work that had a purpose at some point (for certain reorganization) but not anymore.

8.2.3 Explicitly Aligning Training Programs with Actor Roles

An average job description will usually list the tasks, responsibilities and authorities of the person performing the job. However, there are some problems with this.

First of all, a task description does not add anything if the responsibilities have already been clearly defined; on the contrary, a task description makes matters more unclear. A task should coincide exactly with a transaction for which this role is the executor. However, the tasks are rarely so clearly defined that they relate to transactions one-on-one.

Secondly, a job description often not explicitly lists the required competences. Sometimes an ambitious HRM employee, after having had a course on this subject, will point out the importance of competence management, i.e. how employees can acquire and enforce competences.

Sadly when the competences are dealt with, it is seldom assessed to which extent they contribute to the goals of the organization. It frequently happens that HRM just does as it sees fit, however well intended. If such is the case, the connection between the goals of HRM and the corporate goals is missing. This can however be realized once it has been established which actor roles are needed to achieve corporate goals. On the basis of this, it can be clarified which competences are needed in order to perform this actor role optimally. This will allow the HRM department to contribute much more to the strategic interest of the organization.

There is still a lot to be gained in respect of HRM. When I ask employees in an organization to write down what they do exactly and compare this to their job description, these two descriptions turn out to differ so much that it is incomprehensible that nothing has been done about it before. Still, HRM is not to blame, as it is not aware of the underlying cooperation patterns, such as transactions and actor roles. Without that knowledge it is impossible to establish order in the chaos with the usual job- and process descriptions.

8.3 Enterprise Engineering and RACI

Organizations tend to use the so-called RACI model. The RACI model is a matrix that is used to reflect the roles and responsibilities of the persons involved in a project or line duties. At first glance, this method seems very suitable to clearly establish who does what. However, this clarity is not all it seems.

A RACI matrix defines and assigns the following four roles for each task separately:

Responsible—Who is responsible for the task, must ensure that the task is performed correctly?
Accountable—Who is ultimately responsible, must supervise that the task is performed correctly?
Consulted—Who needs to contacted (obligatory or otherwise) for the performance of the task?
Informed—Who needs to be informed on aspects such as progress and result?

Figure 8.1 shows an example of a RACI matrix.

The RACI matrix should make it perfectly clear who does what and thereby provide guidance during the performance of projects or line duties. Employees can use the matrix to look up what their role is for each process and task: the responsible role, the accountable role, the consulted role or the informed role. Unfortunately this description, just as the job descriptions, provides few specifics. Position A will, for example, state that he is responsible for process A, position B is accountable for the same process, function X must be consulted during the process and position Z must be informed. But what does this mean in practice that someone is 'responsible'

RACI table	Employee				
Activity	Wendy	Peter	Remko	Ruud	Phoebe
Define	A	R	I	I	I
Design	I	A	R	C	C
Develop	I	A	R	C	C
Test	A	I	I	R	I

Fig. 8.1 An example of a RACI matrix

or 'accountable'? Is the responsible employee required to obtain a mandate from the accountable for each decision? Or only in some cases? And what information must be sent by whom at which moment to the one who needs to be informed? I can think of many more questions like these. The RACI matrix provides no information on the actual actor roles pertaining to such a RACI role. The exact activities (transactions) still remain vague. Managers and employees benefit much more from real clarity: the design of the construction of an organization with all the transactions and actor roles necessary in order to achieve the intended result. If you have established that, RACI is indeed a very useful tool:

- *Responsible* is the actor role in the performance-dimension who is responsible for the performance of one or more transactions
- *Accountable* is the actor role in the governance-dimension who is responsible for creating the right circumstances for the execution of one or more transactions (in the performance-dimension)
- *Consulted* are the actor roles providing advice on the execution of a transaction for the person who is Responsible
- *Informed* are those actor roles who must know the information that is produced during a transaction (the information bank in a Construction Model)

See Sect. 4.6 for an explanation of the organization dimensions.

8.4 Enterprise Engineering and the Informal Organization

Enterprise Engineering literally 'maps' an organization. The overview that is therewith created uncovers the cooperation within the organization. The interviews and process information will allow the engineer to find out how people work together and therefore how the construction of the formal organization is set up. The current processes are dissected and presented in transactions in a more clear and organized fashion with the aid of the Construction Models. However, something is still missing: the 'informal organization'. The informal organization includes all the transactions that service the self-interest of individual employees rather than the interest of the organization. Self-interest is obviously present in any organization; no organization is free of 'political games'. The greater the uncertainty concerning responsibilities, authorities and competences belonging to certain roles, the greater the chaos in the organization and the more room employees will have to draw up their own plan. Sometimes that 'own plan' will have the purely honorable intention of gaining a grip on matters. In other cases, the 'own plan' will be the result of pure egoism and personal ambition. Someone can, for example, draw attention to himself by taking on a lot of responsibilities, by claiming authorities he doesn't actually have or to perform roles that require competences he does not have. Also hunting for more budget for their own department or making choices that mainly serve his own year-end bonus, are included in the 'informal organization'. The informal organization therefore causes ECFs in the formal organization: the self-interest and the interest of the organization are seldom one and the same.

It is difficult, but not impossible to expose the informal organization in terms of actor roles and transactions. In a direct sense, Enterprise Engineering is a tool to deal with the informal organization. However, I would advise against this. Firstly because the informal organization is rather fleeting; it might be already be changing whilst drawing up the model. This is because interests are fleeting: someone can be your friend today and your threat tomorrow. Secondly, a formal organization set up with Enterprise Engineering allows less room for informal transactions and roles. After all, if the new TO BE model has been implemented correctly, there will be far less chaos. A revised construction is like a restored train track which makes sure that one does not derail and arrives at the right destination: the goals of the organization. If Enterprise prescribes the right cooperation, and the organization monitors such, then the transactions and actor roles that are not in the interest of the organization will stand out quickly and immediate action can be taken.

8.5 Enterprise Engineering and Paying Attention to the People

It should be clear by now that Enterprise Engineering, as focused as it is on the system, does not pay a lot of attention to the somewhat softer management themes, such as personal development. The individual is mainly seen as a link in the system. He 'is' as it were a composite actor role, and his operation is rather analytically dissected in terms of coordination and production activities. This does not mean that personal development of individual employees, an enjoyable work environment and something as inspiring leadership no longer matter. They are simply themes that do not affect the construction and are therefore not relevant for the Enterprise Engineer. They are also themes on which all sorts of statements are given without any kind of scientific substantiation. A nice and free playing field therefore, for anyone who has any ideas about it.

As this chapter and Chap. 6 make clear, Enterprise Engineering is much more than just an improved construction. As we said before, a desired corporate culture can be implemented per coordination activity. It enables an organization that focuses on the wishes of the employees to create extremely substantive tools for a pleasant cooperation. But Enterprise Engineering provides an even bigger contribution to the wellbeing of the employee, all be it indirect.

People often start with a new job full of enthusiasm, and then realize that the activities they are performing have nothing to do with their original expectations. This is partly due to the job description provided in the job advertisement; which is, as we have said before, ambiguous at best. The dissatisfaction may grow quickly, particularly if conflicts arise that are related to ECFs. Someone assumes to be responsible for something, but spends more energy on finding out what he is responsible for exactly. Do you remember the organization monster from the introduction? This phenomenon is also referred to as a communication overhead [1]: the quantity of time people spend on finding out who has which responsibilities, authorities and competences. An overload of meetings that harms productivity, and slowly affects the motivation of the employees. The solution is an organization with a suitable construction and an explicitly aligned HRM department. An employee with a good view of his functioning in terms of the roles he performs, knows exactly what he can expect. Such job description ideally takes into account the production facts that are expected as well. The employee can also effectively work on the growth of his competence, which will allow him to see which competences he should have in order to be able to perform the allocated roles.

Employees generally become happy by doing things they are good at, and become unhappy by the things they are less good at or even bad at. If they are unable to acquire the competences needed for a certain role, it may be time to look out for another position with other underlying actor roles. This means that the grip offered by Enterprise Engineering not only benefits the organization, but also the individual employee, who knows exactly what is expected of him.

8.6 Enterprise Engineering and Lean

In literature, Lean manufacturing, usually abbreviated to Lean, is described as a 'management philosophy'. The aim of Lean is to achieve the maximum value for the customer. This means, realizing a good product delivery with optimum additional conditions, such as delivery time and price-quality ratio. This is to be realized by wasting as little as possible during the manufacturing process. Lean is a philosophy which originated from Japan, and is also referred to as the Toyota Production System (TPS), as it was developed in the Japanese Toyota factories. The process optimization is reached by thoroughly focusing on the processes and to determine, in detail, what can be done faster, cheaper, easier et cetera. Six Sigma is an affiliated management strategy which was developed in the United States. Six Sigma is much more factual than Lean and uses statistical process analyses to detect variations in business processes. Lean and Six Sigma are currently often summarized in one word: Lean.

8.6.1 Positive Points of Lean

By applying Lean a constant focus lies on improving business operations, reducing spillage and creating added value for customers. These all have a positive effect on the business-economical results of businesses.

8.6.2 Lesser Points of Lean

Lean is not a silver bullet, i.e. a solution to all the problems; this much has been made clear by the many millions of cars Toyota has had to recall to the garage these past few years, resulting in a black hole of billions of dollars. The errors that occurred during the manufacturing process (the integration of products to be precise) could apparently not have been prevented by applying TPS/Lean. Lean may be a hype at the moment, but it also has a side that is less good.

Lean involves Kaizen sessions, brainstorm sessions with groups of employees, in which the current processes are discussed to see what can be improved. The highest result achievable during such a session is a consensus on the way in which processes can be improved. It is therefore a consensus model and has little or nothing to do with objectivity, or only by accident. It is impossible to use Lean to check whether hard facts become apparent, and there is no guarantee that such a group process will expose the root cause of a problem. It seems a little naïve to think that asking the 'why-question' five times, which is done during a Kaizen-session, will automatically lead to the root of the problem. If things would be that simple, we would certainly be using this technique in the Middle-East in order to eradicate that hotbed once and for all. If only everything was that simple.

Additionally, the principle of Lean is that processes need to be 'improved'. However, the question of whether these processes are good or complete, or redundant, is not asked; people just assume that they are good. The starting point is the process, but it should be the organization construction with all its Elementary Construction Flaws (ECFs—see Chap. 5). A change initiative can perfectly be facilitated with Lean on the basis of the ECFs. But beware: implementation choices made during the change may lead to new ECFs. An Enterprise Engineer is required to ensure that such does not happen.

Finally there is no theoretical embedding of Lean, in other words, there is no *Way of Thinking* on which Lean is based (see Sect. 5.2). That is why Lean regularly does not have the desired effects in practice, and even unwanted effects. Lean is a tool kit with various tools that may or may not work, and can perfectly be applied in one organization, whilst it may not in other organizations. This brings us back to Sect. 2.1: managing organizations as a black box: we don't know when it works and why this is the case. Probably the lack of a theoretical embedding and the lack of predictable and repeatable results is an important reason why only few scientists in the world have Lean as a study object. However, I do wish to note that Lean does have tools in its tool kit that do have a theoretical embedding. For example Six Sigma, which is based on *Statistical Mathematics*. Another example is the queue theory, which describes, studies and explains characteristics that occur in systems in which customers are required to wait until they are served. This is often applied in logistics (reducing the time spent on logistic processes) and also results from *Statistical Mathematics*. But this method has been around a lot longer than Lean and is an example of a situation that has been occurring more and more frequently: all the means that are somehow relevant to the improvement of processes are placed in Lean's toolkits. Suddenly has become the queue theory as well. This occurrence resembles a black hole; eventually all the means are incorporated in Lean and there will be no escaping the toolkit. When that time comes, the beautifully fat bound book by Bill Bryson 'A short history of nearly everything' will have been reduced to one page and one word: Lean. This does not do justice to all the people who have worked on methods that really add value and I really don't want to see the day that also Enterprise Engineering will be incorporated in Lean's toolkit. It would then suddenly be a hype rather than a means to permanently improve organizations.

Reference

1. Kaufman, J. (2012). *The personal MBA*. London: Portfolio Penguin.

Epilogue

The science of Enterprise Engineering is relatively young but is spreading rapidly throughout scientific circles. In recent years, businesses have started to reap the results of Enterprise Engineering. As is the case with all paradigm changes in science, it will take some time before businesses and society will be able to translate the theory into practice and achieve results with it. In 10 years from now, it may well be possible that we will find it hard to image that we did not use it before. I also expect that in future no curriculum corporate science or organizational science will be taught without the subject Enterprise Engineering.

In this book I mainly looked into the cooperation problems within companies. But the Netherlands, just like any other country, as a whole is an organization too. An organization with actor roles and transactions with which certain goals can be strived for. And the Netherlands has a lot of flaws, here are just a few examples:

A housing cooperation is an organization that focuses on the construction, management and rental or houses (social housing). Since the Dutch government largely gave these organizations the responsibility to build good and cheap houses, and was often only responsible for the financing, regulating and supervising, the housing corporations in the Netherlands have gained a central role in the housing sector. For years, housing corporations have been speculating with their revenues by purchasing risky derivatives. While the housing corporations were not the only owners of these revenues. The Dutch government is part-owner of this money, as they financed the houses. The Dutch government is also supervisor. Without looking into this matter too deeply, we have already established two Elementary Construction Flaws: a housing corporation buys products without having the authority to independently do so and the supervising role of the government has not been performed for years. As things have gone wrong with Vestia (one of the biggest housing cooperations) and others, the government has finally decided to fulfill this role.

Another example is the so-called 'Bulgarians-fraud' which came to light in the spring of 2013. It became apparent that Bulgarians were able to obtain care allowances which they were not entitled to. They did not have the authority to request such an allowance. The municipalities failed to take the responsibility to check whether the applicant had a healthcare insurance. An actor role was therefore

T. Janssen, *Enterprise Engineering*, Management for Professionals,
DOI 10.1007/978-3-319-24172-2

missing in the construction, and the results were disastrous. Many millions disappeared and will never return to the Dutch government.

And finally, in the healthcare sector, there is an ongoing discussion as to whether a patient is a patient or a client. Under the slogan ‘health care institutions can’t manage themselves’ and ‘cost-control’ managers are attracted to ‘deal with this matter once and for all’. But a client is very different from a patient. These are completely different actor roles, with other responsibilities, authorities and competences, and they therefore act totally differently. A client requests a service and it does not matter to him how this service is provided, as long as it is provided. The client also usually doesn’t care whether he or the insurer determines which health care institution will provide the care. But the patient does care. Health care is accompanied by fundamental feelings with regard to trust in the ability of a healthcare institution on the one hand (the one health care provider is better in certain treatments than the other) and the practical aspects on the other hand (you would rather have a health care provider nearby if you don’t have a car). If a patient is regarded a client, the authorities of the patient are limited. The actor role making the request then loses his fundamental right to make an accept, i.e. the right to choose his own health care institutions. This example involves a kind of forced delegation of responsibilities—a purposeful creation of an Elementary Construction Flaw. Another related example: a client may decide to buy an ice cream or not. But the patient has no choice, he needs to be treated. But the government persists, despite all the objections made in society. And in a couple of years the capital error will become visible and things will be reversed to the old situation—and again we will all pay many millions. Just as the Netherlands can be regarded an organization, so can the world as a whole: a system complete with countless Construction Flaws. One example is the different tax systems used by various nations.

Large companies, but also small companies united in a cooperation, pay hardly any taxes due to shrewd legal constructions. This is tax avoidance, and costs around a thousand billion euros a year in Europe alone. This can be regarded a construction flaw. As the tax system differs, there are different rules in different countries. This is inconsistent. The information bank (the rules) should be the same wherever you are, so that national governments wouldn’t miss out on tax revenue as a result of smart detours.

Administrations at international, national and regional levels unintentionally create Construction Flaws. Naturally with the best of intentions. But that does not guarantee success, as the history of mankind has shown us. Construction Flaws can be resolved, but wouldn’t it be much better to do things well straight away, in one go? This would surely save us a lot of trouble. Reorganization after reorganization, centralizing–decentralizing, turning back measures, social unrest due to incomprehensible political decisions, costly programs to force people in a direction they don’t want to, the list is endless. ‘New Way of Working’ (sometimes also referred to as ‘Alternative Workplace Strategies’) is such an example. For reasons of ‘following social change and flexible employment conditions’ cost reductions are introduced in work areas, which for many organizations is the real goal behind the New Way of Working. The buildings are then overpopulated on the assumption that

70 % of the people will work from home. Working from home is perfectly fine, but anyone who enjoys the company of people will experience that the urge to see your colleagues and physically work with them becomes greater. People are social human beings. Whether you like it or not, this is irrelevant for the fact that we enjoy working together and enjoy being meaningful to each other. And we do this according to fixed cooperation patterns which have been incorporated in our genes when communication came into being. The organizations using this fundamental principle will experience that cooperation, both internally and with other parties, will run more smoothly, causing organizations to become more efficient and effective. And it also creates a much more enjoyable work environment. Who wouldn't want that?

MIX
Papier aus verantwortungsvollen Quellen
Paper from responsible sources
FSC® C105338
FSC
www.fsc.org

If you have any concerns about our products,
you can contact us on
ProductSafety@springernature.com

In case Publisher is established outside the EU,
the EU authorized representative is:
Springer Nature Customer Service Center GmbH
Europaplatz 3, 69115 Heidelberg, Germany

Printed by Libri Plureos GmbH
in Hamburg, Germany